T0133200

Austrian Academy of Sciences Press

Austrian Assessment Report Climate Change 2014 (AAR14)

Summary for Policymakers
and
Synthesis

Editors

Helga Kromp-Kolb

Nebojsa Nakicenovic

Karl Steininger

Andreas Gobiet

Herbert Formayer

Angela Köppl

Franz Prettenthaler

Johann Stötter

Jürgen Schneider

The assessment report was produced within the project 'Austrian Panel on Climate Change Assessment Report' funded by 'The Climate and Energy Fund of the Austrian Federal Government' within the framework of the 'Austrian Climate Research Program'.

The present publication is a translation (revised edition) of the Zusammenfassung für Entscheidungstragende and the Synthese from the German language as published by the Austrian Academy of Sciences Press

Vienna, November 2014

ISBN: 978-3-7001-7744-9
ISBN (German version): 978-3-7001-7701-2
ISBN (Complete version) 978-3-7001-7699-2
© with the Authors
© Creative Commons non-commercial 3.0 licence
http://creativecommons.org/icenses/by-nc/3.0/deed.en

The Complete Edition was published with sponsorship of the Austrian Science Fund (FWF): PUB 221-V21

Cover page design
Anka James based on Sabine Tschürtz in Munoz and Steininger, 2010.

Translations from German
Summary for Policymakers: Bano Mehdi
Synthesis: Helga Kromp-Kolb, Nebojsa Nakicenovic, Karl Steininger, Andreas Baumgarten, Birgit Bednar-Friedl, Ulrich Foelsche, Herbert Formayer, Clemens Geitner, Thomas Glade, Andreas Gobiet, Helmut Haberl, Regina Hitzenberger, Martin König, Manfred Lexer, Hanns Moshammer, Klaus Radunsky, Sigrid Stagl, Wolfgang Streicher, Wilfried Winiwarter based on a draft by Adam Pawloff

Copy Editor
Kathryn Platzer, IIASA

Suggested citation of the Summary for Policymakers (SPM)
APCC (2014): Summary for Policymakers (SPM), revised edition. In: Austrian Assessment Report Climate Change 2014 (AAR14), Austrian Panel on Climate Change (APCC), Austrian Academy of Sciences Press, Vienna, Austria.

Suggested citation of the Synthesis
Kromp-Kolb, H., N. Nakicenovic, R. Seidl, K. Steininger, B. Ahrens, I. Auer, A. Baumgarten, B. Bednar-Friedl, J. Eitzinger, U. Foelsche, H. Formayer, C.Geitner, T. Glade, A. Gobiet, G. Grabherr, R. Haas, H. Haberl, L. Haimberger, R. Hitzenberger, M. König, A. Köppl, M. Lexer, W. Loibl, R. Molitor, H.Moshammer, H-P. Nachtnebel, F. Prettenthaler, W.Rabitsch, K. Radunsky, L. Schneider, H. Schnitzer, W.Schöner, N. Schulz, P. Seibert, S. Stagl, R. Steiger, H.Stötter, W. Streicher, W. Winiwarter(2014): Synthesis. In: Austrian Assessment Report Climate Change 2014 (AAR14), Austrian Panel on Climate Change (APCC), Austrian Academy of Sciences Press, Vienna, Austria.

Suggested citation of the complete edition
APCC (2014): Österreichischer Sachstandsbericht Klimawandel 2014 (AAR14). Austrian Panel on Climate Change (APCC), Verlag der Österreichischen Akademie der Wissenschaften, Wien, Österreich, 1096 Seiten. ISBN 978-3-7001-7699-2

This publication includes the Summary for Policymakers (revised edition) and the Synthesis in English. These documents are translated excerpts from the comprehensive descriptions given in the complete edition of the report, to which chapters and volumes cited in this publication refer.

All parts of this report are published in the Austrian Academy of Sciences Press. The full version is available in bookstores.
All publications can be downloaded from www.apcc.ac.at.

Austrian Academy of Sciences Press, Vienna
http://verlag.oeaw.ac.at
http://hw.oeaw.ac.at/7699-2

Print: Wograndl Druck GmbH, 7210 Mattersburg

Printed on acid-free, aging-resistant paper manufactured from non-chlorine bleached pulp
Printed in Austria
http://hw.oeaw.ac.at/7699-2

Austrian Assessment Report Climate Change 2014 (AAR14)
Austrian Panel on Climate Change (APCC)

Project Leader
Nebojsa Nakicenovic

Organizing Committee
Helga Kromp-Kolb, Nebojsa Nakicenovic, Karl Steininger

Project Management
Laura Morawetz

Co-Chairs
Band 1: Andreas Gobiet, Helga Kromp-Kolb
Band 2: Herbert Formayer, Franz Prettenthaler, Johann Stötter
Band 3: Angela Köppl, Nebojsa Nakicenovic, Jürgen Schneider, Karl Steininger

Coordinating Lead Authors
Bodo Ahrens, Ingeborg Auer, Andreas Baumgarten, Birgit Bednar-Friedl, Josef Eitzinger, Ulrich Foelsche, Herbert Formayer, Clemens Geitner, Thomas Glade, Andreas Gobiet, Georg Grabherr, Reinhard Haas, Helmut Haberl, Leopold Haimberger, Regina Hitzenberger, Martin König, Helga Kromp-Kolb, Manfred Lexer, Wolfgang Loibl, Romain Molitor, Hanns Moshammer, Hans-Peter Nachtnebel, Franz Prettenthaler, Wolfgang Rabitsch, Klaus Radunsky, Hans Schnitzer, Wolfgang Schöner, Niels Schulz, Petra Seibert, Sigrid Stagl, Robert Steiger, Johann Stötter, Wolfgang Streicher, Wilfried Winiwarter

Review Editors
Brigitte Bach, Sabine Fuss, Dieter Gerten, Martin Gerzabek, Peter Houben, Carsten Loose, Hermann Lotze-Campen, Fred Luks, Wolfgang Mattes, Sabine McCallum, Urs Neu, Andrea Prutsch, Mathias Rotach

Scientific Advisory Board
Jill Jäger, Daniela Jacob, Dirk Messner

Review Process
Mathis Rogner, Keywan Riahi

Secretariat
Benedikt Becsi, Simon De Stercke, Olivia Koland, Heidrun Leitner, Julian Matzenberger, Bano Mehdi, Pat Wagner, Brigitte Wolkinger

Copy Editing
Kathryn Platzer

Layout and Formatting
Valerie Braun, Kati Heinrich, Tobias Töpfer

Contributing Institutions

The following institutions thankfully enabled their employees to participate in the development of the AAR14 and thus contributed substantially to the report:

- Alpen-Adria University Klagenfurt - Vienna - Graz
- alpS GmbH
- Austrian Academy of Sciences (ÖAW)
- Austrian Agency for Health and Food Safety (AGES)
- Austrian Federal Ministry of Agriculture, Forestry, Environment and Water Management; Dep. IV/4 – Water balance
- Austrian Institute for Technology (AIT)
- Austrian Institute of Economic Research (WIFO)
- Austrian Ministry for Transport, Innovation and Technology, Department for Energy and Environmental Technologies
- Austrian Research Centre for Forests (BFW)
- BIOENERGY2020+ GmbH
- Climate Change Centre Austria (CCCA)
- Climate Policy Initiative, Venice Office
- Danube University of Krems
- Environment Agency Austria
- Federal Agency for Water Management
- Federal Government of Lower Austria
- German Advisory Council on Global Change (WBGU)
- German Development Institute (DIE)
- Graz University of Technology (TU Graz)
- Helmholtz Centre for Environmental Research (UFZ)
- International Institute for Applied Systems Analysis (IIASA)
- J.W.v. Goethe University of Frankfurt am Main
- Joanneum Research Forschungsgesellschaft mbH
- komobile w7 GmbH
- Konrad Lorenz Institute of Ethology
- Lehr- und Forschungszentrum Raumberg-Gumpenstein
- Leibniz Institute for Agricultural Engineering Potsdam-Bornim (ATB)
- Management Center Innsbruck (MCI)
- Max Planck Institute for Meteorology (MPI-M)
- Medical University of Vienna
- Mercator Research Institute on Global Commons and Climate Change
- MODUL University Vienna
- Nature Protection Society Styria
- Office of the Provincial Government of Tyrol
- Potsdam Institute for Climate Impact Research (PIK)
- Society for Renewable Energy Gleisdorf
- Statistics Austria
- Sustainable Europe Research Institute (SERI)
- Swiss Academy of Sciences
- University of Bayreuth
- University of Graz (Uni Graz)
- University of Innsbruck
- University of Leiden
- University of Natural Resources and Life Sciences, Vienna (BOKU)
- University of Salzburg (Uni Salzburg)
- University of Veterinary Medicine Hannover
- University of Vienna (Uni Wien)
- Vienna University of Economics and Business (WU Wien)
- Vienna University of Technology (TU Wien)
- Zentralanstalt für Meteorologie and Geodynamik (ZAMG)

Table of content

Austrian Assessment Report Climate Change 2014 (AAR14)

Foreword

At my inauguration after re-election in 2010, I addressed the challenge of climate change and acknowledged Austria's responsibility to contribute to the solution of this global problem. Since then, in a three-year joint and gratuitous effort, over 200 scientists in Austria have brought together their knowledge across disciplinary boundaries, to jointly paint a comprehensive and scientifically sound picture of climate change in Austria for the public and for decision makers.

Complementary to the global view of the Fifth Assessment Report of the Intergovernmental Panel on Climate Change (IPCC), the Austrian Assessment Report Climate Change (AAR14) of the Austrian Panel on Climate Change (APCC) now summarizes what is known about climate change in Austria, its current and possible future impacts as well as adaptation and mitigation measures. It draws the conclusion that Austria has not sufficiently fulfilled its responsibility to date. But the report also shows that there are many options for action, many of which would be beneficial quite independent of climate change.

The scientific community has impressively demonstrated that they take climate change seriously. Hopefully their work will trigger increased political efforts for climate protection in Austria and strengthen civil society and the wider public in their (growing) engagement for a livable future.

Austrian Assessment Report Climate Change 2014

Summary for Policymakers

Austrian Assessment Report Climate Change 2014

Summary for Policymakers

Coordinating Lead Authors of the Summary for Policymakers
Helga Kromp-Kolb
Nebojsa Nakicenovic
Karl Steininger

Lead Authors of the Summary for Policymakers
Bodo Ahrens, Ingeborg Auer, Andreas Baumgarten, Birgit Bednar-Friedl, Josef Eitzinger, Ulrich Foelsche, Herbert Formayer, Clemens Geitner, Thomas Glade, Andreas Gobiet, Georg Grabherr, Reinhard Haas, Helmut Haberl, Leopold Haimberger, Regina Hitzenberger, Martin König, Angela Köppl, Manfred Lexer, Wolfgang Loibl, Romain Molitor, Hanns Moshammer, Hans-Peter Nachtnebel, Franz Prettenthaler, Wolfgang Rabitsch, Klaus Radunsky, Jürgen Schneider, Hans Schnitzer, Wolfgang Schöner, Niels Schulz, Petra Seibert, Rupert Seidl, Sigrid Stagl, Robert Steiger, Johann Stötter, Wolfgang Streicher, Wilfried Winiwarter

Translation
Bano Mehdi

Citation
APCC (2014): Summary for Policymakers (SPM), revised edition. In: Austrian Assessment Report Climate Change 2014 (AAR14), Austrian Panel on Climate Change (APCC), Austrian Academy of Sciences Press, Vienna, Austria.

Table of content

Introduction

Over the course of a three-year process, Austrian scientists researching in the field of climate change have produced an assessment report on climate change in Austria following the model of the IPCC Assessment Reports. In this extensive work, more than 200 scientists depict the state of knowledge on climate change in Austria and the impacts, mitigation and adaptation strategies, as well as the associated known political, economic and social issues. The Austrian Climate Research Program (ACRP) of the *Klima- und Energiefonds* (Climate and Energy Fund) has enabled this work by financing the coordinating activities and material costs. The extensive and substantial body of work has been carried out gratuitously by the researchers.

This summary for policy makers provides the most significant general statements. First, the climate in Austria in the global context is presented; next the past and future climate is depicted, followed by a summary for Austria on the main consequences and measures. The subsequent section then provides more detail on individual sectors. More extensive explanations can be found – in increasing detail – in the synthesis report and in the full report (Austrian Assessment Report, 2014), both of which are available in bookstores and on the Internet.

The uncertainties are described using the IPCC procedure where three different approaches are provided to express the uncertainties depending on the nature of the available data and on the nature of the assessment of the accuracy and completeness of the current scientific understanding by the authors. For a qualitative evaluation, the uncertainty is described using a two-dimensional scale where a relative assessment is given on the one hand for the quantity and the quality of evidence (i. e. information from theory, observations or models indicating whether an assumption or assertion holds true or is valid), and on the other hand to the degree of agreement in the literature. This approach uses a series of self-explanatory terms such as: high / medium / low evidence, and strong / medium / low agreement. The joint assessment of both of these dimensions is described by a confidence level using five qualifiers from „very high confidence" to „high", „medium", „low" and „very low confidence". By means of expert assessment of the correctness of the underlying data, models or analyses, a **quantitative** evaluation of the uncertainty is provided to assess the likelihood of the uncertainty pertaining to the outcome of the results using eight degrees of probability from „virtually certain" to „more unlikely than likely". The probability refers to the assessment of the likelihood of a well-defined re-

sult which has occurred or will occur in the future. These can be derived from quantitative analyses or from expert opinion. For more detailed information please refer to the Introduction chapter in AAR14. If the description of uncertainty pertains to a whole paragraph, it will be found at the end of it, otherwise the uncertainty assessment is given after the respective statement.

The research on climate change in Austria has received significant support in recent years, driven in particular by the *Klima- und Energiefonds* (Climate and Energy Fund) through the ACRP, the Austrian Science Fund (FWF) and the EU research programs. Also own funding of research institutions has become a major source of funding. However, many questions still remain open. Similar to the process at the international level, a periodic updating of the Austrian Assessment Report would be desirable to enable the public, politicians, administration, company managers and researchers to make the best and most effective decisions pertaining to the long-term horizon based on the most up-to-date knowledge.

The Global Context

With the progress of industrialization, significant changes to the climate can be observed worldwide. For example, in the period since 1880 the global average surface temperature has increased by almost 1 °C. In Austria, this warming was close to 2 °C, half of which has occurred since 1980. These changes are mainly caused by the anthropogenic emissions of greenhouse gases (GHG) and other human activities that affect the radiation balance of the earth. The contribution of natural climate variability to global warming most likely represents less than half of the change. That the increase in global average temperature since 1998 has remained comparatively small is likely attributed to natural climate variability.

Without extensive additional measures to reduce emissions one can expect a global average surface temperature rise of 3–5 °C by 2100 compared to the first decade of the 20th century (see Figure 1). For this increase, self-reinforcing processes (feedback loops), such as the ice-albedo feedback or additional release of greenhouse gases due to the thawing of permafrost in the Arctic regions will play an important role (see Volume 1, Chapter 1; Volume 3, Chapter 1)[1].

---

[1] The full text of the Austrian Assessment Report AAR14 is divided into three volumes, which are further divided into chapters. Information and reference to the relevant section of the AAR14 is provided with the number of the volume (Band) and the respective chapter (Kapitel) where more detailed information can be found pertaining to the summary statements.

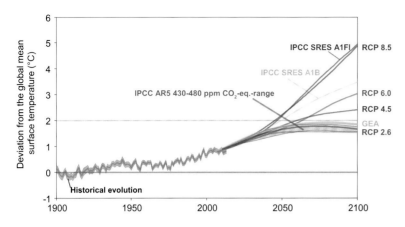

Figure 1 Global mean surface temperature anomalies (°C) relative to the average temperature of the first decade of the 20th century, historical development, and four groups of trends for the future: two IPCC SRES scenarios without emission reductions (A1B and A1F1), which show temperature increases to about 5 °C or just over 3 °C to the year 2100, and four new emission scenarios, which were developed for the IPCC AR5 (RCP8, 5, 6.0, 4.5 and 2.6), 42 GEA emission reduction scenarios and the range of IPCC AR5 scenarios which show the temperature to stabilize in 2100 at a maximum of +2 °C. Data sources: IPCC SRES (Nakicenovic et al. 2000), IPCC WG I (2014) and GEA (2012)

Climate change and the associated impacts show large regional differences. For example, the Mediterranean region can expect a prominent decrease in precipitation as well as associated water availability (see Volume 1, Chapter 4). While, considering the highest emission scenario of a rise in mean sea level of the order 0.5–1 m by the end of the century compared to the current level, poses considerable problems in many densely populated coastal regions (see Volume 1 Chapter 1).

Since the consequences of unbridled anthropogenic climate change would be accordingly serious for humanity, internationally binding agreements on emissions reductions are already in place. In addition, many countries and groups including the United Nations („Sustainable Development Goals"), the European Union, the G-20 as well as cities, local authorities and businesses have set further-reaching goals. In the Copenhagen Accord (UNFCCC Copenhagen Accord) and in the EU Resolution, a goal to limit the global temperature increase to 2 °C compared to pre-industrial times is considered as necessary to limit dangerous climate change impacts. However, the steps taken by the international community on a voluntary basis for emission reduction commitments are not yet sufficient to meet the 2 °C target. In the long-term, an almost complete avoidance of greenhouse gas emissions is required, which means converting the energy supply and the industrial processes, to cease deforestation, and also to change land use and lifestyles (see Volume 3, Chapter 1; Volume 3, Chapter 6).

The likelihood of achieving the 2 °C target is higher if it is possible to achieve a turnaround by 2020 and the global greenhouse gas emissions by 2050 are 30–70 % below the 2010 levels. (see Volume 3, Chapter 1; Volume 3, Chapter 6). Since industrialized countries are responsible for most of the historical emissions – and have benefited from them and hence are also economically more powerful – Article 4 of the UNFCCC suggests that they should contribute to a disproportionate share of total global emission reduction. In the EU „Roadmap for moving to a competitive low-CO_2 economy by 2050" a reduction in greenhouse gas emissions by 80–95 % compared to the 1990 level is foreseen. Despite of the fact that no emission reduction obligations were defined for this period for individual Member States, Austria can expect a reduction commitment of similar magnitude.

Climate Change in Austria: Past and Future

In Austria, the temperature in the period since 1880 rose by nearly 2 °C, compared with a global increase of 0.85 °C. The increased rise is particularly observable for the period after 1980, in which the global increase of about 0.5 °C is in contrast to an increase of approximately 1 °C in Austria (virtually certain, see Volume 1, Chapter 3).

A further temperature increase in Austria is expected (very likely). In the first half of the 21st century, it equals approximately 1.4 °C compared to current temperature, and is not greatly affected by the different emission scenarios due to the inertia in the climate system as well as the longevity of greenhouse gases in the atmosphere. The temperature development thereafter, however, is strongly dependent on anthropogenic greenhouse gas emissions in the years ahead now, and can therefore be steered (very likely, see Volume 1, Chapter 4).

The development of precipitation in the last 150 years shows significant regional differences: In western Austria, an increase in annual precipitation of about 10–15 % was recorded, in the southeast, however, there was a decrease in a similar order of magnitude (see Volume 1, Chapter 3).

In the 21st century, an increase of precipitation in the winter months and a decrease in the summer months is to be expected (likely). The annual average shows no clear trend signal, since Austria lies in the larger transition region between two zones with opposing trends – ranging from an increase in

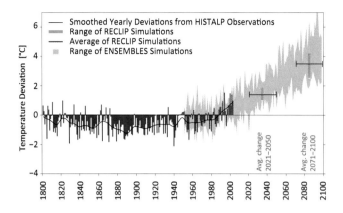

Figure 2 Mean surface air temperature (°C) in Austria from 1800 to 2100, expressed as a deviation from the mean temperature for the period 1971 to 2000. Measurements to the year 2010 are illustrated in color, model calculations for one of the IPCC emissions scenarios with higher GHG emissions (IPCC SRES A1B scenario) in gray. Reproduced are annual means (columns) and the 20-year smoothed curve (line). You can see the temperature drop just before 1900 and the sharp rise in temperature (about 1 °C) since the 1980s. In this scenario, by the end of the century, a rise in temperature of 3.5 °C can be expected (RECLIP simulations). Source: ZAMG

North Europe to a decrease in the Mediterranean (likely, see Volume 1, Chapter 4).

In the last 130 years, the annual sunshine duration has increased for all the stations in the Alps by approximately 20 % or more than 300 hours. The increase in the summer half of the year was stronger than in the winter half of the year (virtually certain, see Volume 1, Chapter 3). Between 1950 and 1980 there was an increase in cloud cover and increased air pollution, especially in the valleys, and therefore a significant decrease in the duration of sunshine hours in the summer (see Volume 1, Chapter 3).

The duration of snow cover has been reduced in recent decades, especially in mid-altitude elevations (approximately 1 000 m above sea level) (very likely, see Volume 2, Chapter 2). Since both the snow line, and thus also the snowpack, as well as the snowmelt are temperature dependent, it is expected that a further increase in temperature will be associated with a decrease in snow cover at mid-altitude elevations (very likely, see Volume 2, Chapter 2).

All observed glaciers in Austria have clearly shown a reduction in surface area and in volume in the period since 1980. For example, in the southern Ötztal Alps, the largest contiguous glacier region of Austria, the glacier area of 144.2 km² in the year 1969 has decreased to 126.6 km² in 1997 and to 116.1 km² in 2006 (virtually certain, see Volume 2, Chapter 2). The Austrian glaciers are particularly sensitive in the retraction phase to summer temperatures since

1980, therefore a further decline of the glacier surface area is expected (very likely). A further increase in the permafrost elevation is expected (very likely, see Volume 2, Chapter 4).

Temperature extremes have changed markedly, so that for example, cold nights are rarer, but hot days have become more common. In the 21st century, this development will intensify and continue, and thus the frequency of heat waves will also increase (very likely, see Volume 1, Chapter 3; Volume 1, Chapter 4,). For extreme precipitation, no uniform trends are detectable as yet (see Volume 1, Chapter 3). However, climate models show that heavy and extreme precipitation events are likely to increase from autumn to spring (see Volume 1, Chapter 4). Despite some exceptional storm events in recent years, a long-term increase in storm activity cannot be detected. Also for the future, no change in storm frequency can be derived (see Volume 1, Chapter 3; Volume 1, Chapter 4).

Summary for Austria: Impacts and Policy Measures

The economic impact of extreme weather events in Austria are already substantial and have been increasing in the last three decades (virtually certain, see Volume 2, Chapter 6). The emergence of damage costs during the last three decades suggests that changes in the frequency and intensity of such damaging events would have significant impacts on the economy of Austria.

The potential economic impacts of the expected climate change in Austria are mainly determined by extreme events and extreme weather periods (medium confidence, see Volume 2, Chapter 6). In addition to extreme events, gradual temperature and precipitation changes also have economic ramifications, such as shifts in potential yields in agriculture, in the energy sector, or in snow-reliability in ski areas with corresponding impacts on winter tourism.

In mountainous regions, significant increases in landslides, mudflows, rockfalls and other gravitational mass movements will occur (very likely, high confidence). This is due to changes in rainfall, thawing permafrost and retreating glaciers, but also to changes in land use (very likely, high confidence). Mountain flanks will be vulnerable to events such as rockfall (very likely, high confidence, see Volume 2, Chapter 4) and landslides (likely, medium confidence, see Volume 2, Chapter 4), and debris masses that were previously fixed by permafrost will be mobilized by debris flows (most likely high confidence, see Volume 2, Chapter 4).

The risk of forest fires will increase in Austria. The risk of forest fires will increase due to the expected warming trend and the increasing likelihood of prolonged summer droughts (very likely, high confidence, see Volume 2, Chapter 4).

Changes to sediment loads in river systems are noticeable. Due to changes in the hydrological and in the sediment regimes (mobilization, transport and deposition) major changes can be expected in mountain torrents and in large river systems (very likely, high confidence, see Volume 2, Chapter 4). The decisive factor here is to distinguish between changes due to climate change and due to human impact.

Due to the currently foreseeable socio-economic development and climate change, the loss potential due to climate change in Austria will increase for the future (medium confidence, see Volume 2, Chapter 3; Volume 2, Chapter 6). A variety of factors determine the future costs of climate change: In addition to the possible change in the distribution of extreme events and gradual climate change, it is mainly socio-economic and demographic factors that will ultimately determine the damage costs. These include, amongst others, the age structure of the population in urban areas, the value of exposed assets, the development of infrastructure for example in avalanche or landslide endangered areas, as well as overall land use, which largely control the vulnerability to climate change.

Without increased efforts to adapt to climate change, Austria's vulnerability to climate change will increase in the decades ahead (high confidence, see Volume 2, Chapter 6). In Austria climate change particularly influences the weather-dependent sectors and areas such as agriculture and forestry, tourism, hydrology, energy, health and transport and the sectors that are linked to these (high confidence, see Volume 2, Chapter 3). It is to be expected that adaptation measures can somewhat mitigate the negative impacts of climate change, but they cannot fully offset them (medium confidence, see Volume 3, Chapter 1).

In 2012 Austria adopted a national adaptation strategy specifically in order to cope with the consequences of climate change (see Volume 3, Chapter 1). The effectiveness of this strategy will be measured principally by how successful individual sectors, or rather policy areas, will be in the development of appropriate adaptation strategies and their implementation. The criteria for their evaluation, such as a regular survey of the effectiveness of adaptation measures, as other nations have already implemented, are not yet developed in Austria.

In 2010 the greenhouse gas emissions in Austria amounted to a total of approximately 81 Mt CO_2-equivalents (CO_2-eq.) or 9.7 t CO_2-eq. per capita (very high confidence, see Volume 1, Chapter 2). These figures take into account the emission contribution of land-use changes through the carbon uptake of ecosystems. The Austrian per capita emis-

sions are slightly higher than the EU average of 8.8 t CO_2-eq. per capita per year and significantly higher than those for example of China (5.6 t CO_2-eq. per capita per year), but much lower than those of the U.S. (18.4 t CO_2-eq. per person per year) (see Volume 1, Chapter 2). Austria has made commitments in the Kyoto Protocol to reduce its emissions. After correcting for the part of the carbon sinks that can be claimed according to the agreement, the emissions for the commitment period 2008 to 2012 were 18.8 % higher than the reduction target of 68.8 M CO_2-eq. per year (see Volume 3, Chapter 1).

By also accounting for the Austrian consumption-related CO_2-emissions abroad, the emission values for Austria are almost 50 % higher (high confidence Volume 3, Chapter 5). Austria is a contributor of emissions in other nations. Incorporating these emissions on the one hand, and adjusting for the Austrian export-attributable emissions on the other hand, one arrives at the „consumption-based" emissions of Austria. These are significantly higher than the emissions reported in the previous paragraph, and in the UN statistics reported for Austria, and this tendency is increasing (in 1997 they were 38 % and in 2004 they were 44 % higher than those reported). From the commodity flows it can be inferred that Austrian imports are responsible for emissions particularly from south Asia and from east Asia, specifically China, and from Russia (see Figure 3).

The national greenhouse gas emissions have increased since 1990, although under the Kyoto Protocol Austria has committed to a reduction of 13 % over the period 2008 to 2012 compared to 1990 (virtually certain, see Volume 3, Chapter 1; Volume 3, Chapter 6). The Austrian goal was set relatively high compared to other industrialized countries. Formally compliance with this reduction target for 2008 to 2012 was achieved through the purchase of emission rights abroad amounting to a total of about 80 Mt CO_2-eq. for roughly € 500 million (very high confidence, see Volume 3, Chapter 1).

In Austria, efforts are underway to improve energy efficiency and to promote renewable energy sources; however, the objectives pertaining to renewables and energy efficiency are not sufficiently backed by tangible measures to make them achievable. Thus, in 2010 an energy strategy was released which proposes that the final energy consumption in 2020 should not exceed the level of 2005; an amount of 1 100 PJ. However, this has not yet been implemented with adequate measures. Austria's Green Electricity Act (*Ökostromgesetz*) stipulates that an additional power generation of 10.5 TWh (37.8 PJ) per year up to 2020 should be from renewable sources. The energy

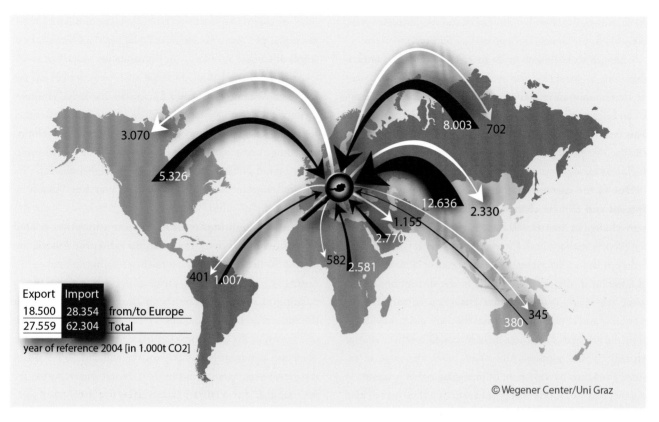

Export	Import	
18.500	28.354	from/to Europe
27.559	62.304	Total

year of reference 2004 [in 1.000t CO2]

© Wegener Center/Uni Graz

Figure 3 CO$_2$ streams from the trade of goods to/from Austria according to major world regions. The emissions implicitly contained in the imported goods are shown with red arrows, the emissions contained in the exported goods, attributed to Austria, are shown with white arrows. Overall, south Asia and east Asia, particularly China, and Russia, are evident as regions from which Austria imports emission-intensive consumer- and capital- goods. Source: Munoz and Steininger (2010)

sector and the industry are largely regulated under the „EU ETS", the further development of which is currently negotiated. In particular, the transport sector currently lacks effective measures.

Austria has set only short-term reduction targets for its climate and energy program, namely for the period up to 2020 (see Volume 3, Chapter 1; Volume 3, Chapter 6). This corresponds to the binding EU targets, but to adequately tackle the problem other countries have set longer-term GHG reduction targets. For example, Germany has set a reduction target of 85 % to 2050. The UK intends to achieve a reduction of 80 % by 2050 (see Volume 3, Chapter 1).

The measures taken so far are insufficient to meet the expected contribution of Austria to achieve the global 2 °C target (high confidence, see Volume 3, Chapter 1; Volume 3, Chapter 6). The actions specified by Austria are based on the objectives for the year 2020; the goals for developing renewable energy sources in Austria are not sufficiently ambitious and are likely to be achieved well before 2020. It is unlikely that an actual change in emission trends will be achieved in the industrial and transport sectors, while the turnaround that

has already taken place for space heating is likely to be insufficient (see Volume 3, Chapter 3; Volume 3, Chapter 5). The expected greenhouse gas emissions savings due to the replacement of fossil fuels with biofuels are increasingly being called into question (see Volume 3, Chapter 2).

Institutional, economic, social and knowledge barriers slow progress with respect to mitigation and adaptation. Measures to eliminate or overcome these barriers include a reforming of administrative structures with respect to relevant tasks at hand, such as the pricing of products and services according to their climate impact. A key factor in this regard includes an abolition of environmentally harmful financing and subsidies; for example, for the exploration of new fossil reserves, or the commuter support which favors the use of the cars, or housing subsidies for single-family homes in the urban vicinity. Also, having a strong involvement of civil society and of science in the decision-making processes can accelerate necessary measures. Relevant knowledge gaps should be addressed because they also delay further action, however they do not belong to the most important factors (high confidence, see Volume 3, Chapter 1; Volume 3, Chapter 6).

According to scenario simulations, emission reductions of up to 90 % can be achieved in Austria by 2050 through additional implementation measures (high confidence, see Volume 3, Chapter 3; Volume 3, Chapter 6). These scenarios are obtained from studies that focus on the energy supply and demand. However, currently there is a lack of clear commitment on the part of the decision-makers to emission reductions of such a magnitude. In addition, so far there is no clear perception pertaining to the financial or other economic and social framework conditions on how the listed objectives could be achieved. In addition to technological innovations, far-reaching economic and socio-cultural changes are required (e. g. in production, consumption and lifestyle).

According to the scenarios, the target set by the EU can be achieved by halving the energy consumption in Austria by 2050. It is expected that the remaining energy demand can be covered by renewable energy sources. The economically available potential of renewable resources within Austria is quantified at approximately 600 PJ. As a comparison, the current final energy consumption is 1 100 PJ per year (see Volume 3, Chapter 3). The potential to improve energy efficiency exists, particularly in the sectors of buildings, transportation and production (high confidence, see Volume 3, Chapter 3; Volume 3, Chapter 5).

Striving for a swift and serious transformation to a carbon-neutral economic system requires a cross-sectoral closely coordinated approach with new types of institutional cooperation in an inclusive climate policy. The individual climate mitigation strategies in the various economic sectors and related areas are not sufficient. Other types of transformations should also be taken into account, such as those of the energy system, because decentralized production, storage and control system for fluctuating energy sources and international trade are gaining in importance (medium confidence, see Volume 3, Chapter 3). Concurrently, numerous small plant operators with partially new business models are entering the market.

An integrative and constructive climate policy contributes to managing other current challenges. One example is economic structures become more resistant with respect to outside influences (financial crisis, energy dependence). This means the intensification of local business cycles, the reduction of international dependencies and a much higher productivity of all resources, especially of energy (see Volume 3, Chapter 1).

The achievement of the 2050 targets only appears likely with a paradigm shift in the prevailing consumption and behavior patterns and in the traditional short-term ori-

ented policies and decision-making processes (high confidence, see Volume 3, Chapter 6). Sustainable development approaches which contribute both to a drastic departure from historical trends as well as individual sector-oriented strategies and business models can contribute to the required GHG reductions (probably, see Volume 3, Chapter 6). New integrative approaches in terms of sustainable development require not necessarily novel technological solutions, but most importantly a conscious reorientation of established, harmful lifestyle habits and in the behavior of economic stakeholders. Worldwide, there are initiatives for transformations in the direction of sustainable development paths, such as the energy turnaround in Germany (*Energiewende*), the UN initiative „Sustainable Energy for All", a number of „Transition Towns" or the „Slow Food" movement and the vegetarian diet. Only the future will show which initiatives will be successful (see Volume 3, Chapter 6).

Demand-side measures such as changes in diet, regulations and reduction of food losses will play a key role in climate protection. Shifting to a diet based on dominant regional and seasonal plant-based products, with a significant reduction in the consumption of animal products can make a significant contribution to greenhouse gas reduction (most likely, high confidence). The reduction of losses in the entire food life cycle (production and consumption) can make a significant contribution to greenhouse gas reduction. (very likely, medium confidence).

The necessary changes required to attain the targets include the transformation of economic organizational forms and orientations (high confidence, see Volume 3, Chapter 6). The housing sector has a high need for renewal; the renovation of buildings can be strengthened through new financing mechanisms. The fragmented transport system can be further developed into an integrated mobility system. In terms of production, new products, processes and materials can be developed that also ensure Austria is not left behind in the global competition. The energy system can be aligned along the perspective of energy services in an integrated manner.

In a suitable political framework, the transformation can be promoted (high confidence, see Volume 3, Chapter 1; Volume 3, Chapter 6). In Austria, there is a willingness to change. Pioneers (individuals, businesses, municipalities, regions) are implementing their ideas already, for example in the field of energy services, or climate-friendly mobility and local supply. Such initiatives can be strengthened through policies that create a supportive environment.

New business and financing models are essential elements of the transformation. Financing instruments (beyond the subsidies primarily used so far) and new business models relate mainly to the conversion of the energy selling enterprises to specialists for energy services. The energy efficiency can be significantly increased and made profitable, legal obligations can drive building restoration, collective investments in renewables or efficiency measures can be made possible by adapting legal provisions. Communication policy and regional planning can facilitate the use of public transport and emission-free transport, such as is the case for example in Switzerland (see Volume 3, Chapter 6). Long-term financing models (for buildings for example for 30 to 40 years), which are especially endowed by pension funds and insurance companies can facilitate new infrastructure. The required transformation has global dimensions, therefore efforts abroad, showing solidarity, should be discussed, including provisions for the Framework Convention Climate Fund.

Major investments in infrastructure with long lifespans limit the degrees of freedom in the transformation to sustainability if greenhouse gas emissions and adaptation to climate change are not considered. If all projects had a „climate-proofing" subject to consider integrated climate change mitigation and appropriate adaptation strategies, this would avoid so-called „lock-in effects" that create long-term emission-intensive path dependencies (high confidence, see Volume 3, Chapter 6). The construction of coal power plants is an example. At the national level this includes the disproportionate weight given to road expansion, the construction of buildings, which do not meet current ecological standards – that could be met at justifiable costs – and regional planning with high land consumption inducing excessive traffic.

A key area of transformation is related to cities and densely settled areas (high confidence, see Volume 3, Chapter 6). The potential synergies in urban areas that can be used in many cases to protect the climate are attracting greater attention. These include, for example, efficient cooling and heating of buildings, shorter routes and more efficient implementation of public transport, easier access to training or education and thus accelerated social transformation.

Climate-relevant transformation is often directly related to health improvements and accompanied by an increase in the quality of life (high confidence, see Volume 3, Chapter 4; Volume 3, Chapter 6). For the change from car to bike, for example, a positive-preventive impact on cardiovascular diseases has been proven, as have been further health-improving effects, that significantly increase life expectancy, in addition to positive environmental impacts. Health supporting effects have also been proven for a sustainable diet (e. g. reduced meat consumption).

Climate change will increase the migration pressure, also towards Austria. Migration has many underlying causes. In the southern hemisphere, climate change will have particularly strong impacts and will be a reason for increased migration mainly within the Global South. The IPCC estimates that by 2020 in Africa and Asia alone 74 million to 250 million people will be affected. Due to the African continent being particularly impacted, refugees from Africa to Europe are expected to increase (Volume 3, Chapter 4).

Climate change is only one of many global challenges, but a very central one (very high confidence, see Volume 2, Chapter 6; Volume 3, Chapter 1; Volume 3, Chapter 5). A sustainable future also deals with for example issues of combating poverty, a focus on health, social human resources, the availability of water and food, having intact soils, the quality of the air, loss of biodiversity, as with ocean acidification and overfishing (very high confidence, see Volume 3, Chapter 6). These questions are not independent of each other: climate change often exacerbates the other problems. And therefore it often affects the most vulnerable populations the most severely. The community of states has triggered a UN process to formulate sustainable development goals after 2015 (Sustainable Development Goals). Climate change is at the heart of these targets and many global potential conflict areas. Climate mitigation measures can thus generate a number of additional benefits to achieve further global objectives (high confidence, see Volume 3, Chapter 6).

Impacts on Sectors and Measures of Mitigation and Adaptation

Soils and Agriculture

Climate change leads to the loss of humus and to greenhouse gas emissions from the soil. Temperature rise, temperature extremes and dry periods, more pronounced freezing and thawing in winter as well as strong and long drying out of the soil followed by heavy precipitation enhance certain processes in the soil that can lead to an impairment of soil functions, such as soil fertility, water and nutrient storage capacity, humus depletion causing soil erosion, and others. This results in increased greenhouse gas emissions from soil (very likely, see Volume 2, Chapter 5).

Human intervention increases the area of soils with a lower resilience to climate change. Soil sealing and the consequences of unsuitable land use and management such as compaction, erosion and loss of humus further restrict soil functions and reduce the soil's ability to buffer the effects of climate change (very likely, see Volume 2, Chapter 5).

The impacts of climate change on agriculture vary by region. In cooler, wetter areas – for example, in the northern foothills of the Alps – a warmer climate mainly increases the average potential yield of crops. In precipitation poorer areas north of the Danube and in eastern and south-eastern Austria, increasing drought and heat-stress reduce the long term average yield potential, especially of non-irrigated crops, and increase the risk of failure. The production potential of warmth-loving crops, such as corn or grapes, will expand significantly (very likely, see Volume 2, Chapter 3).

Heat tolerant pests will propagate in Austria. The damage potential of agriculture through – in part newly emerging – heat tolerant insects will increase. Climate change will also alter the occurrence of diseases and weeds (very likely, see Volume 2, Chapter 3).

Livestock will also suffer from climate change. Increasing heat waves can reduce the performance and increase the risk of disease in farm animals (very likely, see Volume 2, Chapter 3).

Adaptation measures in the agricultural sector can be implemented at varying rates. Within a few years measures such as improved evapotranspiration control on crop land (e.g. efficient mulch cover, reduced tillage, wind protection), more efficient irrigation methods, cultivation of drought- or heat-resistant species or varieties, heat protection in animal husbandry, a change in cultivation and processing periods as well as crop rotation, frost protection, hail protection and risk insurance are feasible (very likely, see Volume 3, Chapter 2).

In the medium term, feasible adaptation measures include soil and erosion protection, humus build up in the soil, soil conservation practices, water retention strategies, improvement of irrigation infrastructure and equipment, warning, monitoring and forecasting systems for weather-related risks, breeding stress-resistant varieties, risk distribution through diversification, increase in storage capacity as well as animal breeding and adjustments to stable equipment and to farming technology (very likely, see Volume 3, Chapter 2).

The shifts caused by a future climate in the suitability for the cultivation of warmth-loving crops (such as grain corn, sunflower, soybean) is shown in Figure 4 for the example of grapes for wine production. Many other heat tolerant crops such as corn, sunflower or soybean show similar expansions in areas suitable for their cultivation in future climate as is shown here for the case of wine (see Volume 2, Chapter 3).

Agriculture can reduce greenhouse gas emissions in a variety of ways and enhance carbon sinks. If remaining at current production volume levels, the greatest potentials lie in the areas of ruminant nutrition, fertilization practices, reduction of nitrogen losses and increasing the nitrogen efficiency (very likely, see Volume 3, Chapter 2). Sustainable strategies for reducing greenhouse gas emissions in agriculture require resource-saving and efficient management practices involving organic farming, precision farming and plant breeding whilst conserving genetic diversity (probably, see Volume 3, Chapter 2).

Forestry

A warmer and drier climate will strongly impact the biomass productivity of Austrian forests. Due to global warming, the biomass productivity increases in mountainous areas and in regions that receive sufficient precipitation. However, in eastern and northeastern lowlands and in inner-alpine basins, the productivity declines, due to more dry periods (high agreement, robust evidence, see Volume 2, Chapter 3; Volume 3, Chapter 2).

In all of the examined climate scenarios, the disturbances to forest ecosystems are increasing in intensity and in frequency. This is particularly true for the occurrence of heat-tolerant insects such as the bark beetle. In addition, new types of damage can be expected from harmful organisms that have been imported or that have migrated from southern regions. Abiotic disturbances such as storms, late and early frosts, wet snow events or wildfires could also cause greater damages than before (high uncertainty). These disturbances can also trigger outbreaks and epidemics of major forest pests, such as the bark beetle. Disturbances lead to lower revenues for wood production. The protective function of the forests against events such as rockfalls, landslides, avalanches as well as carbon storage decrease (high agreement, robust evidence, see Volume 2, Chapter 2; Volume 3, Chapter 2).

For decades Austria's forests have been a significant net sink for CO_2. Since approximately 2003, the net CO_2 uptake of the forest has declined and in some years has come to a complete standstill; this is due to higher timber harvests, natural disturbances and other factors. In addition to the GHG impacts of increased felling, a comprehensive greenhouse gas balance of different types of forest management and use of forest products requires considering the carbon storage in long-lived wood products as well as the GHG savings of other emission-intensive products that can be replaced by wood (e.g. fossil fuel, steel, concrete) as well. A final assessment of the systemic effects would require more accurate and comprehensive analyzes than those that currently exist (see Volume 3, Chapter 2).

The resilience of forests to risk factors as well as the adaptability of forests can be increased. Examples of ad-

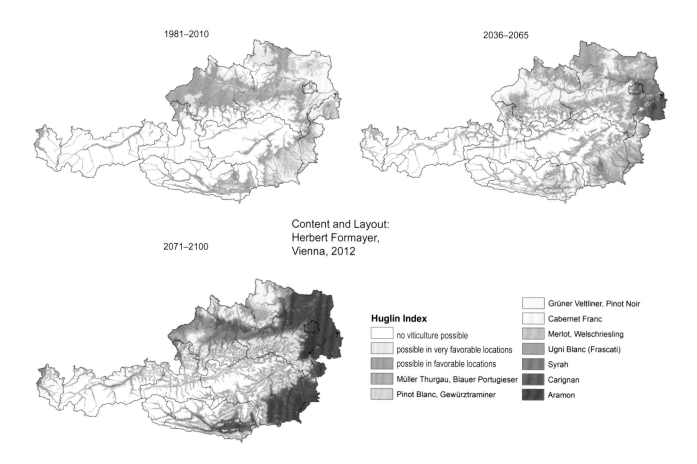

1981–2010

2036–2065

2071–2100

Content and Layout:
Herbert Formayer,
Vienna, 2012

Huglin Index

- no viticulture possible
- possible in very favorable locations
- possible in favorable locations
- Müller Thurgau, Blauer Portugieser
- Pinot Blanc, Gewürztraminer
- Grüner Veltliner, Pinot Noir
- Cabernet Franc
- Merlot, Welschriesling
- Ugni Blanc (Frascati)
- Syrah
- Carignan
- Aramon

Figure 4 Evolution of the climatic suitability for the cultivation of different varieties, taking into account the optimum heat levels and rainfall in Austria in the past climate (observed) and a climate scenario until the end of the 21st century (modelled). The color shades from blue to yellow to purple indicate increasing heat amounts exclusively based on the corresponding variety classification. One can clearly see the increasing suitability for red wines, towards the end of the century as there are extremely heat-loving varieties. Source: Eitzinger and Formayer (2012)

aptation measures are smaller scale management structures, mixed stands adapted to sites, and ensuring the natural forest regeneration in protected forests through adapted game species management. The most sensitive areas are the spruce stands in mixed deciduous forest sites located in lowlands, and spruce monocultures in mountain forests serving a protective function. The adaptation measures in the forest sector are associated with considerable lead times (high agreement, robust evidence, see Volume 3, Chapter 2).

Biodiversity

Ecosystems that require a long time to develop, as well as alpine habitats located above the treeline are particularly impacted by climate change (high agreement, robust evidence, see Volume 2, Chapter 3). Bogs and mature forests require a long time to adapt to climate change and are therefore particularly vulnerable. Little is known about the interaction with other elements of global change, such as land use change or the introduction of invasive species. The adaptive capacity of species and habitats has also not been sufficiently researched.

In alpine regions, cold-adapted plants can advance to greater heights and increase the biodiversity in these regions. Cold-adapted species can survive in isolated microniches in spite of the warming (high agreement, robust evidence). However, increasing fragmentation of populations can lead to local extinctions. High mountains native species that have adapted to lower peripheral regions of the Alps are particularly affected (medium agreement, medium evidence, see Volume 2, Chapter 3).

Animals are also severely affected. In the animal kingdom, changes in the annual cycles are already documented, such as the extension of activity periods, increased successions of generations, earlier arrival of migratory birds, as well as shifts in distribution ranges northward or to higher elevations of individual species. Climate change will further advantageous for some animal species, especially generalists, and fur-

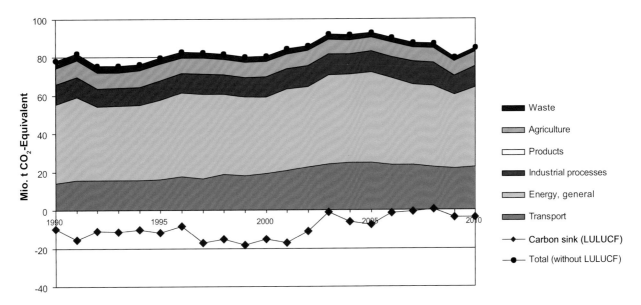

Figure 5 Officially reported greenhouse gas emissions in Austria (according to the IPCC source sectors with especially defined emissions for the Transport sector). The brown line that is mainly below the zero line represents carbon sinks. The sector „Land use and land use change" (LULUCF) represents a sink for carbon and is therefore depicted below the zero line. In recent years, this sink was significantly smaller and no longer present in some years. This was mainly a result of higher felling; and changes to the survey methods contributed to this as well. Source: Anderl et al. (2012)

ther endanger others, especially specialists (medium evidence, see Volume 2, Chapter 3). The warming of rivers and streams leads to a theoretical shift in the fish habitat by up to 30 km. For brown trout and grayling for example, the number of suitable habitats will decline (high agreement, robust evidence, see Volume 2, Chapter 3).

Energy

Austria has a great need to catch up on improvements in energy intensity. In the last two decades, unlike the EU average, Austria has made little progress in terms of improvements to energy intensity (energy consumption per GDP in Euro, see Figure 6). Since 1990, the energy intensity of the EU-28 decreased by 29 % (in the Netherlands by 23 %, Germany by 30 % and in the UK by 39 %). In Germany and the UK, however some of these improvements are due to the relocation of energy-intensive production abroad. In terms of emission intensity (GHG emissions per PJ energy) the improvements in Austria since 1990 are a reflection of the strong development of renewables; here, Austria along with The Netherlands, counts among the countries with the strongest improvements. These two indicators together determine the greenhouse gas emission intensity of the gross domestic product (GDP), which in Austria as well as in the EU-28 has also declined since 1990. Greenhouse gas emissions have increased more

slowly than GDP. However, in comparison with the EU-28 it becomes evident that Austria must make major strides to catch up in reducing energy intensity (see Volume 3, Chapter 1).

The potential renewable energy sources in Austria are currently not fully exploited. In Austria, the share of renewable energy sources in the gross final energy consumption has increased from 23.8 % to 31 % between 2005 and 2011, primarily due to the development of biogenic fuels, such as pellets and biofuels. In the future, wind and photovoltaics can make a significant contribution. The target for 2020, for a 34 % share in end energy use of renewable energies can be easily achieved with the current growth rates. However, for the required medium-term conversion to a greenhouse gas neutral energy system by 2050, a coverage of the entire energy demand with renewable energy sources is necessary. To avoid a mere shifting of the problem, before any further future expansion of hydroelectric power or increased use of biomass takes place, it is important to examine the total greenhouse gas balances as well as to take into account indirect and systemic effects. Other environmental objectives do not lose their importance in an effort to protect the climate (see Volume 3, Chapter 3; Volume 3, Chapter 6).

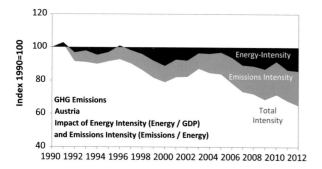
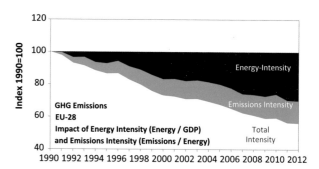
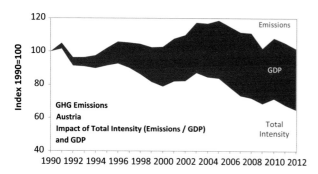
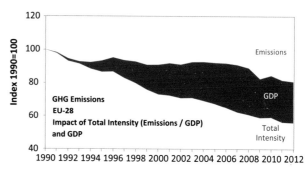

Figure 6 Development of GHG intensity of GDP and the subdevelopments of energy intensity (energy consumption per GDP in Euro) and emission intensity of energy (greenhouse gas emissions per PJ of energy) over time for Austria and for the EU-28 (upper panel). The development of greenhouse gas emission intensity in conjunction with rising GDP (lower panel) leads to rising greenhouse gas emissions for Austria (+5%), and declining emissions for the EU-28 (−18%) during this period; Source: Schleicher (2014)

Transport and Industry

Of all sectors, the greenhouse gas emissions increased the most in the last two decades in the transport sector by +55% (very high confidence, see Volume 3, Chapter 3). Efficiency gains made in vehicles were largely offset by heavier and more powerful vehicles as well as higher transport performance. However, the limitations of CO_2 emissions per kilometer driven for passenger cars and vans are beginning to bear fruit (see Volume 3, Chapter 3). Public transportat supply changes and (tangible) price signals have had demonstrable effects on the share of private vehicle transport in Austria.

To achieve a significant reduction in greenhouse gas emissions from passenger transport, a comprehensive package of measures is necessary. Keys to achieving this are marked reductions in the use of fossil-fuel energy sources, increasing energy efficiency and changing user behaviour. A prerequisite is improved economic- and settlement- structures in which the distances to travel are minimized. This may strengthen the environmentally friendly forms of mobility used, such as walking and cycling. Public transportation systems are to be expanded and improved, and their CO_2 emissions are to be minimized. Technical measures for car transport include further, massive improvements in efficiency for vehicles or the use of alterna-

tive power sources (Volume 3, Chapter 3) – provided that the necessary energy is also produced with low emissions.

Freight transportation in Austria, measured in tonne-kilometers, increased faster in the last decades than the gross domestic product. The further development of transport demand can be shaped by a number of economic and social conditions. Emissions can be reduced by optimizing the logistics and strengthening the CO_2 efficiency of transport. A reduction in greenhouse gas emissions per tonne-kilometer can be achieved by alternative power and fuels, efficiency improvements and a shift to rail transportation (see Volume 3, Chapter 3).

The industry sector is the largest emitter of greenhouse gases in Austria. In 2010, the share of the manufacturing sector's contribution to the total Austrian energy-consumption as well as to greenhouse gas emissions was almost 30%, in both cases. Emission reductions in the extent of about 50% or more cannot be achieved within the sector through continuous, gradual improvements and application of the relevant state of the art of technology. Rather, the development of climate-friendly new procedures is necessary (radical new technologies and products with a drastic reduction of energy consumption), or the necessary implementation of procedures for the storage of the greenhouse gas emissions (carbon capture and storage,

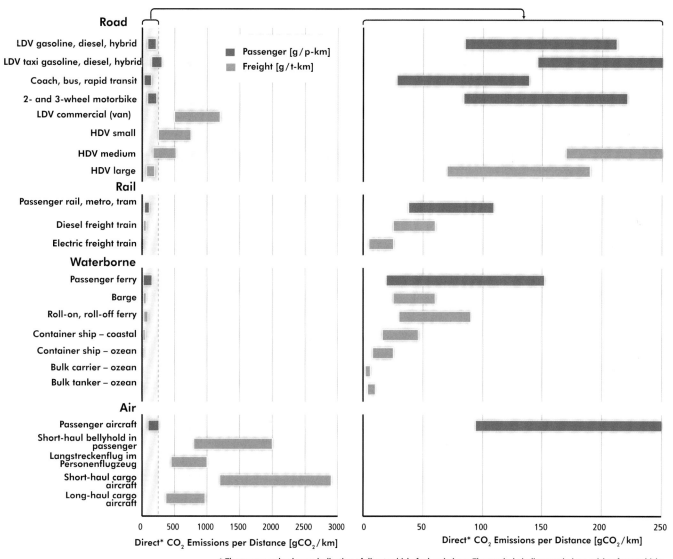

Figure 7 A comparison of characteristic CO_2 emissions per passenger-kilometer and ton-kilometer for different transport modes that use fossil energy and thermal electricity generation in case of electric railways. Source: IPCC (2014)

for example as in the EU scenarios for Energy Roadmap 2050) (very likely, see Volume 3, Chapter 5).

Tourism

Winter tourism will come under pressure due to the steady rise in temperature. Compared to destinations where natural snow remains plentiful, many Austrian ski areas are threatened by the increasing costs of snowmaking (very likely, see Volume 3, Chapter 4).

Future adaptation possibilities with artificial snow-making are limited. Although currently 67 % of the slope surfaces are equipped with snowmaking machines, the use of these is limited by the rising temperatures and the (limited) availability of water (likely, see Volume 3, Chapter 4). The promotion of the development of artificial snow by the public sector could therefore lead to maladaptation and counterproductive lock-in effects.

Tourism could benefit in Austria due to the future very high temperatures expected in summer, in the Mediterra-

nean (very likely). However, even with equally good turnout and capacity utilization in the summer, the value added lost in winter cannot be regained with an equal gain in visitor numbers in summer (see Volume 3, Chapter 4).

Losses in tourism in rural areas have high regional economic follow-up costs, since the loss of jobs often cannot be compensated by other industries. In peripheral rural areas, which already face major challenges due to the demographic change and the increasing wave of urbanization, this can lead to further resettlement (see Volume 3, Chapter 1; Volume 3, Chapter 4).

Urban tourism may experience set-backs in midsummer due to hot days and tropical nights (very likely). Displacements of the stream of tourists in different seasons and regions are possible and currently already observable (see Volume 3, Chapter 4).

Successful pioneers in sustainable tourism are showing ways to reduce greenhouse gases in this sector. In Austria there are flagship projects at all levels – individuals, municipalities and regions – and in different areas, such as hotels, mobility, and lucrative offers for tourists. Due to the long-term investment in infrastructure for tourism, lock-in effects are particularly vulnerable (see Volume 3, Chapter 4).

Infrastructure

Energy use for heating and cooling buildings and their GHG emissions can be significantly reduced (high agreement, see Volume 3, Chapter 5). A part of this potential can be realized in a cost-effective manner. To further reduce the energy demand of existing buildings, high-quality thermal renovation is necessary. For energy supply, mainly alternative energy sources, such as solar thermal or photovoltaic are to be used for the reduction of greenhouse gas emissions. Heat pumps can only be used in the context of an integrated approach which ensures low CO_2 power generation, thereby contributing to climate protection. Biomass will also be important in the medium term. District heating and cooling will become less important in the long term due to reduced demand. A significant contribution to future greenhouse gas neutrality in buildings can also be provided by building construction standards, which the (almost) zero-energy and plus-energy houses promote. These are foreseen to occur across the EU after 2020. Given the large number of innovative pilot projects, Austria could assume a leadership role in this area also before. Targeted construction standards and renovation measures could significantly reduce the future cooling loads. Specific zonal planning and building regulations can ensure denser designs with higher

energy efficiency, especially also beyond the inner urban settlement areas (see Volume 3, Chapter 5).

Forward planning of infrastructure with a long service life under changing conditions can avoid poor investments. Against the background of continuously changing post-fossil energy supply conditions, infrastructure projects in urban locations, in transport and energy supplies should be reviewed to ensure their emission-reducing impacts as well as their resilience to climate change. The structure of urban developments can be designed so that transport and energy infrastructures are coordinated and built (and used) efficiently with low resource consumptions (see Volume 3, Chapter 5).

A decentralized energy supply system with renewable energy requires new infrastructure. In addition to novel renewables with stand-alone solutions (e. g. off-grid photovoltaics) there are also new options for integrating these onto the network. Local distribution networks for locally produced biogas as well as networks for exploiting local, mostly industrial, waste heat (see Volume 3, Chapter 1; Volume 3, Chapter 3) require special structures and control. „Smart Grids" and „Smart Meters" enable locally produced energy (which is fed into the grid, e. g. from co- and poly-generation or private photovoltaic systems) to contribute to improved energy efficiency and are therefore discussed as elements of a future energy system (see Volume 3, Chapter 5). However, there are concerns of ensuring network security as well as data protection and privacy protection; these issues are not yet sufficiently defined or regulated by law.

Extreme events can increasingly impair energy and transport infrastructures. Longer duration and more intense heat waves are problematic (very likely), more intense rainfall and resulting landslides and floods (probably), storms (possible) and increased wet-snow loads (possible, see Volume 1, Chapter 3; Volume 1, Chapter 4; Volume 1, Chapter 5; Volume 2, Chapter 4) pose potential risks for infrastructure related to settlement, transportation, energy and communications. If an increase in climate damages and costs are to be avoided, the construction and expansion of urban areas and infrastructure in areas (regions) that are already affected by natural hazards should be avoided. Moreover, when designating hazard zones, the future development in the context of climate change should be taken as a precautionary measure. Existing facilities can provide increased protection through a range of adaptation measures, such as the creation of increased retention areas against flooding.

The diverse impacts of climate change on water resources require extensive and integrative adaptation measures.

Both high- and low-water events in Austrian rivers can negatively impact several sectors, from the shipping industry, the provision of industrial and cooling water, to the drinking water supply. The drinking water supply can contribute to adaptation measures through the networking of smaller supply units as well as the creation of a reserve capacity for source water (high agreement, robust evidence, see Volume 3, Chapter 2).

Adaptation measures to climate change can have positive ramifications in other areas. The objectives of flood protection and biodiversity conservation can be combined through the protection and expansion of retention areas, such as floodplains (high agreement, much evidence). The increase in the proportion of soil organic matter leads to an increase in the soil water storage capacity (high agreement, robust evidence, see Volume 2, Chapter 6) and thus contributes to both flood protection and carbon sequestration, and therefore to climate protection (see Volume 3, Chapter 2).

Health and Society

Climate change may cause directly- or indirectly- related problems for human health. Heat waves can lead to cardiovascular problems, especially in older people, but also in infants or the chronically ill. There exists a regional-dependent temperature at which the death rate is determined to be the lowest; beyond this temperature the mortality increases by 1–6 % for every 1 °C increase in temperature (very likely, high confidence, see Volume 2, Chapter 6; Volume 3, Chapter 4). In particular, older people and young children have shown a significant increase in the risk of death above this optimum temperature. Injuries and illnesses that are associated with extreme events (e. g. floods and landslides) and allergies triggered by plants that were previously only indigenous to Austria, such as ragweed, also add to the impacts of climate change on health.

The indirect impacts of climate change on human health remains a major challenge for the health system. In particular, pathogens that are transferred by blood-sucking insects (and ticks) play an important role, as not only the agents themselves, but also the vectors' (insects and ticks) activity and distribution are dependent on climatic conditions. Newly introduced pathogens (viruses, bacteria and parasites, but also allergenic plants and fungi such as, e. g. ragweed (*Ambrosia artemisiifolia*) and the oak processionary moth (*Thaumetopoea processionea*)) and new vectors (e. g., „tiger mosquito", *Stegomyia albopicta*) can establish themselves, or existing pathogens can spread regionally (or even disappear). Such imported cases are virtually unpredictable and the opportunities to take counter-measures are low (likely, medium confidence, see Volume 2, Chapter 6).

Health-related adaptations affect a myriad of changes to individual behavior of either a majority of the population or by members of certain risk groups (likely, medium agreement, see Volume 3, Chapter 4). Several measures of adaptation and mitigation that are not primarily aimed at improving human health may have significant indirect health-related benefits, such as switching from a car to a bike (likely, medium agreement, see Volume 3, Chapter 4).

The health sector is both an agent and a victim of climate change. The infrastructure related to the health sector requires both mitigation and adaptation measures. Effective mitigation measures could include encouraging the mobility of employees and patients as well as in the procurement of used and recycled products (very likely, high agreement, see Volume 3, Chapter 4). For specific adaptation to longer-term changes there is a lack of medical and climate research, however some measures can be taken now – such as in preparing for heat waves.

Vulnerable groups generally are more highly exposed to the impacts of climate change. Usually the confluence of several factors (low income, low education level, low social capital, precarious working and living conditions, unemployment, limited possibilities to take action) make the less privileged population groups more vulnerable to climate change impacts. The various social groups are affected differently by a changing climate, thus the options to adapt are also dissimilar and are also influenced by differing climate policy measures (such as higher taxes and fees on energy) (likely, high agreement, see Volume 2, Chapter 6)

Climate change adaptation and mitigation lead to increased competition for resource space. This mainly affects natural and agricultural land uses. Areas for implementing renewable energy sources, or retention areas and levees to reduce flood risks are often privileged at the expense of agricultural land. Increasing threats of natural hazards to residential areas may lead to more resettlements in the long term (high confidence, see Volume 2, Chapter 2; Volume 2, Chapter 5). In order to facilitate the adaptation of endangered species to climate change by allowing them to migrate to more suitable locations and in order to better preserve biodiversity, conservation areas must be drawn up and networked with corridors (high confidence, see Volume 3, Chapter 2). There is no regional planning strategy for Austria that can provide necessary guidelines for relevant decisions (see Volume 3, Chapter 6).

Transformation

Although in all sectors significant emission reduction potentials exist, the expected Austrian contribution towards achieving the global 2 °C target cannot be achieved with sector-based, mostly technology-oriented, measures. Meeting the 2 °C target requires more than incrementally improved production technologies, greener consumer goods and a policy that (marginal) increases efficiency to be implemented in Austria. A transformation is required concerning the interaction of the economy, society and the environment, which is supported by behavioral changes of individuals, however these changes also have to originate from the individuals. If the risk of unwanted, irreversible change should not increase, the transformation needs to be introduced and implemented rapidly (see Volume 3, Chapter 6).

A transformation of Austria into a low-carbon society requires partially radical structural and technical renovations, social and technological innovation and participatory planning processes (medium agreement, medium evidence, see Volume 3, Chapter 6). This implies experimentation and experiential learning, the willingness to take risks and to accept that some innovations will fail. Renewal from the root will be necessary, also with regards to the goods and services that are produced by the Austrian economy, and large-scale investment programs. In the assessment of new technologies and social developments an orientation along a variety of criteria is required (multi-criteria approach) as well, an integrative socio-ecologically oriented decision-making is needed instead of short-term, narrowly defined cost-benefit calculations. To be of best effectiveness, national action should be agreed upon internationally, both with the surrounding nations as well as with the global community, and particularly in partnership with developing countries (see Volume 3, Chapter 6).

In Austria, a socio-ecological transformation conducive to changes in people's belief-systems can be noticed. Individual pioneers of change are already implementing these ideas with climate-friendly action and business models (e. g. energy service companies in real estate, climate-friendly mobility, or local supply) and transforming municipalities and regions (high agreement, robust evidence). At the political level, climate-friendly transformation approaches can also be identified. If Austria wants to contribute to the achievement of the global 2 °C target and help shape a future climate-friendly development at a European level and internationally, such initiatives need to be reinforced and supported by accompanying policy measures that create a reliable regulatory landscape (high agreement, medium evidence, see Volume 3, Chapter 6).

Policy initiatives in climate mitigation and adaptation are necessary at all levels in Austria if the above objectives are to be achieved: at the federal level, at that of provinces and that of local communities. Within the federal Austrian structure the competences are split, such that only a common and mutually adjusted approach across those levels can ensure highest effectiveness and achievement of objectives (high agreement; strong evidence). For an effective implementation of the – for an achievement necessarily – substantial transformation a package drawing from the broad spectrum of instruments appears to be the only appropriate one (high agreement, medium evidence).

Figure Credits

Figure 1 Issued for the AAR14 adapted from: IPCC, 2013: In: Climate Change 2013: The Physical Science Basis. Contribution of Working Group I to the Fifth Assessment Report of the Intergovernmental Panel on Climate Change [Stocker, T.F., D. Qin, G.-K. Plattner, M. Tignor,S. K. Allen, J. Boschung, A. Nauels, Y. Xia, V. Bex and P.M. Midgley (Eds.)]. Cambridge University Press, Cambridge, United Kingdom and New York, NY, USA.; IPCC, 2000: Special Report on Emissions Scenarios [Nebojsa Nakicenovic and Rob Swart (Eds.)]. Cambridge University Press, UK.; GEA, 2012: Global Energy Assessment - Toward a Sustainable Future, Cambridge University Press, Cambridge, UK and New York, NY, USA and the International Institute for Applied Systems Analysis, Laxenburg, Austria.

Figure 2 Issued for the AAR14 adapted from: Auer, I., Böhm, R., Jurkovic, A., Lipa, W., Orlik, A., Potzmann, R., Schöner, W., Ungersböck, M., Matulla, C., Briffa, K., Jones, P., Efthymiadis, D., Brunetti, M., Nanni, T., Maugeri, M., Mercalli, L., Mestre, O., Moisselin, J.-M., Begert, M., Müller-Westermeier, G., Kveton, V., Bochnicek, O., Stastny, P., Lapin, M., Szalai, S., Szentimrey, T., Cegnar, T., Dolinar, M., Gajic-Capka, M., Zaninovic, K., Majstorovic, Z., Nieplova, E., 2007. HISTALP – historical instrumental climatological surface time series of the Greater Alpine Region. International Journal of Climatology 27, 17–46. doi:10.1002/joc.1377; ENSEMBLES project: Funded by the European Commission's 6th Framework Programme through contract GOCE-CT-2003-505539; reclip:century: Funded by the Austrian Climate Research Program (ACRP), Project number A760437

Figure 3 Muñoz, P., Steininger, K.W., 2010: Austria's CO_2 responsibility and the carbon content of its international trade. Ecological Economics 69, 2003–2019. doi:10.1016/j.ecolecon.2010.05.017

Figure 4 Issued for the AAR14. Source: ZAMG

Figure 5 Anderl M., Freudenschuß A., Friedrich A., et al., 2012: Austria's national inventory report 2012. Submission under the United Nations Framework Convention on Climate Change and under the Kyoto Protocol. REP-0381, Wien. ISBN: 978-3-99004-184-0

Figure 6 Schleicher, St., 2014: Tracing the decline of EU GHG emissions. Impacts of structural changes of the energy system and economic activity. Policy Brief. Wegener Center for Climate and Global Change, Graz. Basierend auf Daten des statistischen Amtes der Europäischen Union (Eurostat)

Figure 7 ADEME, 2007; US DoT, 2010; Der Boer et al., 2011; NTM, 2012; WBCSD, 2012, In Sims R., R. Schaeffer, F. Creutzig, X. Cruz-Núñez, M. D'Agosto, D. Dimitriu, M.J. Figueroa Meza, L. Fulton, S. Kobayashi, O. Lah, A. McKinnon, P. Newman, M. Ouyang, J.J. Schauer, D. Sperling, and G. Tiwari, 2014: Transport. In: Climate Change 2014: Mitigation of Climate Change. Contribution of Working Group III to the Fifth Assessment Report of the Intergovernmental Panel on Climate Change [Edenhofer, O., R. Pichs-Madruga, Y. Sokona, E. Farahani, S. Kadner, K. Seyboth, A. Adler, I. Baum, S. Brunner, P. Eickemeier, B. Kriemann, J. Savolainen, S. Schlömer, C. von Stechow, T. Zwickel and J.C. Minx (Eds.)]. Cambridge University Press, Cambridge, United Kingdom and New York, NY, USA

Austrian Assessment Report Climate Change 2014 (AAR14)

Synthesis

Austrian Assessment Report Climate Change 2014 (AAR14)

Synthesis

Coordinating Lead Authors of the Synthesis

Helga Kromp-Kolb
Nebojsa Nakicenovic
Rupert Seidl
Karl Steininger

Lead Authors of the Synthesis

Bodo Ahrens, Ingeborg Auer, Andreas Baumgarten, Birgit Bednar-Friedl, Josef Eitzinger, Ulrich Foelsche, Herbert Formayer, Clemens Geitner, Thomas Glade, Andreas Gobiet, Georg Grabherr, Reinhard Haas, Helmut Haberl, Leopold Haimberger, Regina Hitzenberger, Martin König, Angela Köppl, Manfred Lexer, Wolfgang Loibl, Romain Molitor, Hanns Moshammer, Hans-Peter Nachtnebel, Franz Prettenthaler, Wolfgang Rabitsch, Klaus Radunsky, Jürgen Schneider, Hans Schnitzer, Wolfgang Schöner, Niels Schulz, Petra Seibert, Sigrid Stagl, Robert Steiger, Johann Stötter, Wolfgang Streicher, Wilfried Winiwarter

Citation

Kromp-Kolb, H., N. Nakicenovic, R. Seidl, K. Steininger, B. Ahrens, I. Auer, A. Baumgarten, B. Bednar-Friedl, J. Eitzinger, U. Foelsche, H. Formayer, C. Geitner, T. Glade, A. Gobiet, G. Grabherr, R. Haas, H. Haberl, L. Haimberger, R. Hitzenberger, M. König, A. Köppl, M. Lexer, W. Loibl, R. Molitor, H. Moshammer, H-P. Nachtnebel, F. Prettenthaler, W. Rabitsch, K. Radunsky, L. Schneider, H. Schnitzer, W. Schöner, N. Schulz, P. Seibert, S. Stagl, R. Steiger, H. Stötter, W. Streicher, W. Winiwarter (2014): Synthesis. In: Austrian Assessment Report Climate Change 2014 (AAR14), Austrian Panel on Climate Change (APCC), Austrian Academy of Sciences Press, Vienna, Austria.

Table of content

S.0   Introduction

S.0.1   Motivation

The Austrian Assessment Report 2014 (AAR14) was conceived as a national counterpart to the periodically compiled assessment reports of the Intergovernmental Panel on Climate Change (IPCC). Whereas the IPCC reports focus on the global and regional levels, the AAR14 focuses on the situation in Austria. AAR14, which follows, was compiled by Austrian scientists working in the field of climate change over a three-year period and was modelled on the IPCC assessment report process. In this extensive publication, more than 200 scientists depict the state of knowledge on climate change in Austria: the impacts, mitigation, and adaptation strategies, as well as the associated political, economic, and social issues. AAR14 presents a coherent and consistent report i) on historically observed climate change and its impacts on the environment and society; and ii) on potential future trends and options in the areas of adaptation and mitigation in Austria In so doing, it takes into account country-specific natural, societal, and economic characteristics. It provides much needed knowledge about regional manifestations of global climate change. The report also indicates gaps in knowledge and understanding. Like the IPCC reports, AAR14 is based on contributions that have already been published and aims to be policy-relevant without being policy-prescriptive.

The Austrian Climate Research Program (ACRP) of the *Klima- und Energiefonds* (Climate and Energy Fund) has financed the coordinating activities and material costs of this study. The extensive and substantial body of work has been carried out gratuitously by the researchers, with strong support of their respective institutions.

This synthesis is divided into three sections corresponding to the three volumes of the full report. It provides the most significant information from the full report based on the contributions of the coordinating lead-authors of the individual chapters of the AAR14. The sections are as follows:

- **Volume 1**. Climate change in Austria: Drivers and Manifestations (coordinating lead-author: Helga Kromp-Kolb).

Volume 1 describes the scientific basis of climate change, in particular its historical and future manifestations in Austria.

- **Volume 2.** Climate change in Austria: Environmental and Societal Implications (coordinating lead-author: Rupert Seidl)

Volume 2 describes the impacts of climate change on the hydro-, bio, pedo- and lithospheres and on humans, the economy, and society (anthroposphere).

- **Volume 3**. Climate change in Austria: Mitigation and Adaptation (coordinating lead-authors: Nebojsa Nakicenovic and Karl Steininger)

Volume 3 introduces options to mitigate greenhouse gas (GHG) emissions and to adapt to climate change. Possible transformation paths toward a more climate-friendly society and economy are presented.

References to individual chapters within this synthesis are made by reference to the volume and chapter of the report itself (e.g., Volume 1, Chapter 3)[1]. References to the original literature can be found in the main report. Cited "grey literature" is available from the literature database of the Climate Change Centre Austria (CCCA) (www.ccca.ac.at).

S.0.2   Handling Uncertainties; Safety and Precautionary Principle

All insights, even scientific insights, are subject to uncertainty. In the public debate on climate change, uncertainty has often been used to justify postponing decisions and actions. From a scientific point of view, however, uncertainty must be properly dealt with. This report shows that it is possible to take decisions on the basis of existing knowledge, despite uncertainty.

Uncertainty regarding the scientific reliability of the theory of anthropogenic climate change (in the following: climate change theory) is nurtured by media and popular scientific books and films, which offer a broad spectrum of alternative interpretations. Epistemologically speaking, strict proof of climate change theory is impossible (Volume 1, Chapter 5), quite apart from the fact that a prediction of the future is impossible. However, the theory of human-induced climate change is well supported by model experiments and empirical studies and, furthermore, has been subject to scientific scrutiny for over 40 years. As a result, the mainstream climate change theory is superior to all other theories and hypotheses that have thus far been presented. As long as no new evidence or insights emerge that challenge the core of climate change theory, it is appropriate to base societal, political and economic decisions upon it.

---

[1]   The 1092 page report is available only in German, with English headings and captions of figures and tables.

Within climate change theory, the reliability of individual statements varies. For example, most statements regarding future temperature changes are more robust than statements regarding future levels of precipitation. Uncertainty can arise for a variety of reasons, such as lack of data, lack of understanding of processes, or the lack of a generally accepted explanation for observations or model results.

The IPCC has developed a specific terminology to express uncertainties that uses three different approaches. The choice of approach depends on the nature of the available data and on the authors' assessment of the accuracy and completeness of the current scientific understanding. For a **qualitative** estimation, the uncertainty is described using a two-dimensional scale where a relative assessment is given, on the one hand, for the quantity and the quality of evidence (i.e., information from theory, observations, or models indicating whether an assumption or assertion holds true or is valid) and, on the other hand, to the degree of agreement in the literature. This approach uses a series of self-explanatory terms such as high / medium / low evidence and strong / medium / low agreement. The joint assessment of both these dimensions is described by a confidence level, using five qualifiers: "very high," "high," "medium," "low," and "very low." For a **quantitative assessment**, expert judgment of the correctness of the underlying data, models, or analyses, is used to assess the uncertainty of the results, using eight degrees of probability from "virtually certain" to "more unlikely than likely." The probability is based on the assessment of the likelihood of a well-defined result having occurred or being expected to occur in the future. The degrees of probability can be derived from quantitative analyses or from expert opinion. For more detailed information, please refer to the introductory chapter in AAR14 (in German) or the relevant IPCC documents (in English). In the following, if the description of uncertainty relates to a whole paragraph, it will be found at the end of that paragraph. Otherwise, the uncertainty assessment is given after the statement in question.

As the report deals with both past and future developments, uncertainty allows for the fact that the future will be influenced by human activity. In climate and climate impact research, this is typically dealt with by applying different scenarios in which various potential future developments are presented, without actually developing prognoses.

The selection of scenarios is not limited to the most likely developments, as climate change is just as much an ethical issue as an academic one. From an ethical point of view, not at all impacts of climate change are equally important (Volume 1, Chapter 5). From an ethical perspective, it is especially important to study impacts that risk violating basic human rights, such as the right to life, health, and autonomy. It is generally accepted that future generations have rights that must be respected by people alive today. This principle – in the form of "sustainable development" – has been on the international agenda since the "Brundtland Report" (1987) and can thus be considered as a fundamental and internationally recognized ethical consensus. As a result, climate policy that unnecessarily puts people's fundamental rights at risk is impermissible.

Principles of environmental ethics, which are integrated in differing ways into the body of law of many countries, provide an orientation: the security principle, the precautionary principle, and the polluter-pays principle. In cases of uncertainty, the security principle demands that upper limits (worst-case scenarios) of possible negative environmental impacts are assumed. Implementing climate protection measures does not require scientific proof of negative climate change impacts beyond doubt; a plausible and justified suspicion is sufficient. Consequently, doubts about the anthropogenic nature of climate change are no justification for business as usual. For climate science this means that for society to take informed decisions, the full bandwidth of possible impacts must be depicted, including potential best-case and worst-case scenarios, even if they are unlikely.

S.0.3 Acknowledgments

Over 250 people worked on AAR14. They are acknowledged in the full report, and they delivered the basis for this synthesis. At this juncture it is only possible to express deepest thanks collectively to all authors, lead authors, coordinating lead-authors, co-chairs, reviewers, review editors, members of the quality management and scientific advisory boards, the project manager, the secretariat, the lectors, and those who did the page layout. Thanks also go to all institutions that made this report possible, either through financial or in-kind contributions, and to the Climate and Energy Fund and the FWF Austrian Science Fund for their financial support.

S.1 Climate Change in Austria: Drivers and Manifestations

S.1.1 The global climate system and causes of climate change

The progress of industrialization has caused significant observable changes to the climate worldwide. For example, in the period since 1880 the global average surface temperature has in-

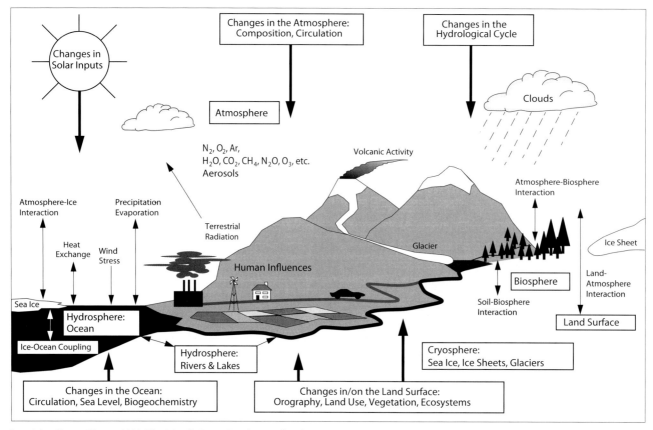

Figure S.1.1. Graphical overview over climate subsystems (boxes, bold font), their exchanges (thin arrows, normal font) and some aspects which change (thick arrows). The most relevant trace gases and aerosols are mentioned. Source: Houghton et al. (2001)

creased by almost 1 °C. An understanding of the causes of these changes is a prerequisite for estimating possible future changes.

The climate system can be considered as an externally influenced, dynamic system, the state of which "changes" at time scales of years to geological eras. The climate is influenced by subsystems such as the atmosphere, hydrosphere, and biosphere. These spheres store and exchange energy, water, carbon, and trace elements (Figure S.1.1). Such processes can frequently be represented as cycles. The sun delivers the energy to sustain all the (climate) processes on earth. Solar energy enters the climate system as solar radiation. A large part of the solar radiation, which is absorbed by the earth's surface, is emitted back into the atmosphere as terrestrial radiation, where it is partly absorbed and then radiated back again to earth, manifesting itself as the greenhouse effect in the climate system. The terrestrial radiation that is not absorbed by the atmosphere is radiated back into space. A relatively small amount of the energy absorbed by the earth is taken up by the biosphere, for example, through photosynthesis. When the climate system is in global balance, solar (incoming) radiation (wavelength 0,3–3

μm) and terrestrial (outgoing) radiation (3–100 μm) balance each other out over several years.

If terrestrial radiation decreases, for example, through an increase in carbon dioxide (CO_2), nitrous oxide (N_2O), methane (CH_4), ozone (O_3), chlorofluorocarbons, sulphur hexafluoride (SF_6), or water vapor (H_2O), the net energy intake of the climate system can increase.

Beside the greenhouse effect, there are three major factors that influence the energy exchange between the earth and outer space and thereby influence radiative forcing and the average surface temperature of the planet:

- The radiant flux from the sun reaching the earth. This is subject to natural fluctuations; however, the latter have not amounted to more than 0.5 W/m² in the past 400 years, which is quite minimal when compared to the average value of 1 361 W/m².

- Changes in the parameters of the earth's orbit, at scales of several hundred to several hundred thousand years (Milankovic theory).

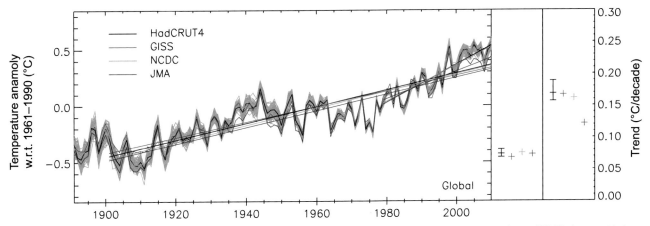

Figure S.1.2. Time-series of global surface temperature anomalies (reference period 1961 to 1990) with uncertainty bounds, calculated by four international research groups. Trends on the right are calculated for 1900 to 2010 and 1980 to 2010, and are statistically highly significant. Source: Morice et al. (2012)

- The planetary albedo, the share of incoming solar radiation that is reflected by the earth and its atmosphere without absorption. The albedo is determined by clouds, amount and distribution of snow and ice, aerosol particles in the atmosphere, and the type of land cover and land use. Changes in the albedo of the magnitude of merely 1 % have a significant influence on net radiation.

The water budget, like the energy budget, also plays a central role. Water vapor is the most important greenhouse gas (GHG); however, anthropogenic emissions of water vapor are insignificant when compared to natural evaporation. With rising temperatures, the content of water vapor in the atmosphere increases, which, because of increased long-wave atmospheric radiation, leads to a positive feedback and increases the warming caused by longer-lived GHGs. Water vapor is often not included in GHG budgets due to its short life time in the troposphere and the comparatively small amount of direct anthropogenic emissions.

To explain the observed increase in GHGs in the atmosphere, it is necessary to consider biogeochemical cycles, in particular the carbon budget, which includes processes such as photosynthesis, respiration, storage and respiration in oceans, and anthropogenic activities. Anthropogenic sources are causing an increase in atmospheric CO_2 content; this is leading to an increase in natural sink activity, in particular enhanced photosynthesis (increased biomass production) and stronger uptake of CO_2 in the oceans (ocean acidification).

Although the specific influences of human beings on the climate system are very complex, the majority of observed changes in climate since 1880 can be explained by just a few activities:

1. The combustion of fossil fuels (coal, oil, gas) and the related increase in GHG emissions.
2. Land use change (e. g., deforestation, afforestation, soil sealing) and agriculture (e. g., deforestation, sealing, nitrogen fertilizer, humus decomposition, methane emissions from rice fields and the stomachs of ruminants).
3. Process-related emissions from industry (e. g., production of cement and steel).

The most important source of GHGs in the last 50 years has been the combustion of fossil fuels, which has tripled during this time. Although natural CO_2 sinks have increased along with the increase in the CO_2 concentration in the atmosphere, they cannot offset rising anthropogenic CO_2 emissions. The most recent figures estimate current (2011) anthropogenic CO_2 emissions at 10.4 ±1.1 Gt C/year, of which 9.5 ±0.5 Gt C/year can be attributed to the combustion of fossil fuels and cement production and 0.9 ±0.6 Gt C/year to land use change. Of the anthropogenic emissions, 2.5 ±0.5 Gt C/year are absorbed by the oceans and 2.6 ±0.8 Gt C/year by the terrestrial biosphere, whereas 4.3 ±0.1 Gt C/year remain in the atmosphere. Accordingly, the CO2 content in the atmosphere has increased by approximately 30 % since 1959.

This increase, which is easily measurable, is one of the most important foundations of the insight that anthropogenic CO_2 emissions lead to an increase in CO_2 concentration. The cumulative anthropogenic CO_2 emissions since 1870 are approximately 1479 Gt CO_2 (400 Gt C). The carbon content of the atmosphere has risen by 840 Gt CO_2 (or by 230 Gt C, which is an increase of 39% over pre-industrial levels). The concentration of the second most important GHG, methane, has doubled since 1870. The fifth IPCC report, published in 2013, estimates the contribution of all anthropogenic GHGs to radiative forcing at 1.9 W/m² ±1 W/m².

Although current climate change is most apparent through the increase in mean global temperature, it is also revealed through a number of other parameters such as distribution of precipitation and shifting of climate zones. In essence, a pole-ward shifting of climate zones and an enlargement of arid environments can be observed. Changes in the cryosphere (all forms of snow and ice) are also dramatic. These changes relate not only to glacial melting in the Alps and other mountains but also to the melting of the Greenland ice sheet and the reduction of Arctic sea ice in summer. The thermal expansion of oceans and the melting of land-based glaciers and ice sheets are leading to a rise in sea level, increasingly endangering coastal regions: between 1880 and 2009 the sea level rose by a global average of around 20 cm.

Past climates, before the instrumental period, can be reconstructed using proxy data, for example, fossils or deposits from past geological epochs. Past temperatures during these periods can be particularly inferred from isotope ratios in deep sea sediments and from ice cores. For the Holocene, the time following the last cold period, a number of other proxy data are available, such as tree rings, pollen, and corals, to name a few.

The climate of the current geological period of the past 2.5 million years, the Quaternary (Pleistocene and Holocene) has been an interplay of long glacials, with mean global temperatures as much as 6°C below current values, and short interglacials (warm periods) with temperatures similar to today, driven by variations in earth orbit parameters (shape of the orbit, tilt, and orientation of the rotation axis of the earth). Within this period we currently live in a warm period. The Holocene, which started about 11700 years ago, was characterized by a relatively stable climate. During the last 2000 years there have been warmer periods worth mentioning (in around 1000 A.D.) and colder periods (in the 17th century and around 1850).

Global temperatures have been rising since around 1850, and both proxy and instrumental data show these increases tending to accelerate during the past decades. The rate of warming in the last decades of the 20th century was particularly dramatic compared to climate variations during the Holocene. The rapid increase of approximately 1°C observed in the last 100 years (see Figure S.1.2) is not extreme from a geological

perspective; it is, however, the first time that an increase has been caused by anthropogenic activity, and it is the beginning of anticipated, considerably stronger warming.

The anthropogenic influence on the current climate can be determined on the basis of available observations, the modelled reconstruction of the past (re-analyses), elaborate statistical methods (so-called fingerprint methods), and climate simulations. The direct conclusion from this evidence is that future climatic changes will be significantly influenced by global socioeconomic developments. In this context many different trajectories are conceivable; these depend on parameters that are hard to forecast, such as population and economic growth, the use and development of emission mitigation technologies, availability of resources, and political decisions. In other words, future climatic developments will also depend on human decisions.

For the Fifth IPCC Assessment Report (IPCC AR5), four so-called Representative Concentration Pathways (RCPs) were developed, which provide the basis for climate projections. The individual pathways all include different trends in GHG emissions that lead to radiative forcing values between 2.6 (RCP 2.6) and 8.5 W/m² (RCP 8.5) by 2100 (Volume 1, Chapter 1), which in turn all lead to a stabilization of radiative forcing at different levels and within varying time frames.

Using earth system simulation models (advanced global climate models), parameters such as temperature, pressure, and precipitation changes are calculated on the basis of RCP emission pathways. Mean global surface temperature provides a general description of the anthropogenic warming of the earth's atmosphere. It is both a symbol and a valuable indicator for overall climate change. Figure S.1.3 demonstrates that the internationally agreed political goal of limiting warming to a maximum of 2°C relative to preindustrial temperature levels can be reached only in the most ambitious concentration pathway (RCP 2.6) In RCP 2.6 global radiative forcing levels reach a maximum before 2050, in RCP 4.5 are stabilized after around 2080, and in RCP 6.0 after around 2150. However, temperature still increases after these points in time, due to the inertia of the climate system and in particular of the oceans. Temperature differences between the pathways only become significant around the middle of the 21st century and after.

As many regional climate studies and almost all climate impact studies are still based on the IPCC SRES-scenarios used prior to the IPCC AR5, reference will be made to these repeatedly in the following pages. If the increases in temperature by the end of the century are compared, then the extreme RCP 8.5 approximately equates to the SRES A1F1 scenario, RCP 6.0 to SRES B2, and RCP 4.5 to SRES B1. The often used

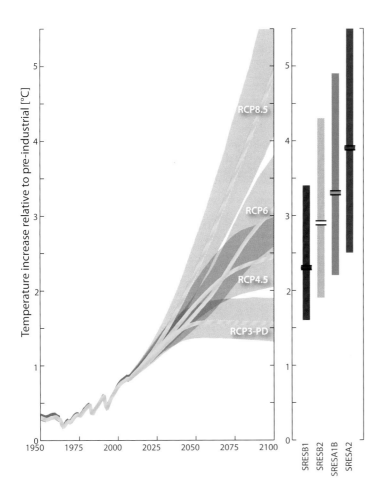

Figure S.1.3. Observed and simulated global average temperatures near the surface for the period 1950–2100, shown as deviations from the mean temperature of 1980–1999, for four representative concentration pathways (RCPs). Source: Rogelj et al. (2012)

SRES A2 scenario is in the region of RCP 8.5. A SRES scenario similar to RCP 2.6 and equivalent to the 2 °C target was not part of the SRES group. In accordance with IPCC specifications at the time, SRES scenarios did not take account of mitigation activities nor, therefore, of stabilization. The range of possible developments in the 21st century foreseen within the new RCPs is therefore broader than that of the SRES scenarios.

S.1.2 Emissions, Sinks, and Concentrations of Greenhouse Gases and Aerosols

In 2010 Austrian greenhouse gas emissions totaled approximately 81 Mt CO_2-equivalents (81 000 Gg CO_2-eq.),[2, 3] that

---

[2] 1 Gg = 109 g, equals to 1 kt (thousand tonnes) and 1 Tg = 1 012 g = 1 Mt (million tonnes)

[3] All references to total GHG emissions consider the respective "global warming potential" (GWP). GWP describes the global warming potential of a substance over a period of 100 years in relation to CO_2. In this way GHGs can be converted into CO_2-equivalents and be considered in their sum. According to this definition the GWP of CO_2 is equal to 1. In this report, mandatory GWP values for report-

---

is, around 0.17 % of global emissions.[4] At 9.7 t CO_2-eq., Austrian annual per capita emissions are slightly higher than the EU annual average of 8.8 t CO_2-eq. per capita, considerably higher than Switzerland's at 6.9 t CO_2-eq., but significantly lower than those of the USA (18.4 t CO_2-eq.). Although Austria committed itself under the Kyoto Protocol to reduce GHG emissions by 13 % between 1990 and 2010, its 2010 emissions – if decreasing carbon sinks are taken into account – were around 19 % above 1990 levels (see Figure S.1.4).

Fossil fuel use causes the largest share of Austrian national GHG emissions, almost 63 Mt CO_2-emissions 2010 (78 % of total national GHG emissions). Over 17 % of emissions are attributed to energy conversion (power stations, refineries, coke ovens), almost 20 % to industrial energy conversion, approximately 13 % to heating (9 % of which in private households),

---

ing to UNFCCC (United Nations Framework Convention on Climate Change) are used (IPCC 1996): 21 for CH_4 (i.e., the effect of 1 kg CH_4 is equivalent to 21 kg of CO_2), 310 for N_2O, and between 140 and 23 900 for different fluorinated compounds.

[4] Natural biochemical cycles are not included as they are considered to be a constant background. All emissions data shown refer to 2010.

and the majority of the remaining emissions (over 27%) to transport – all percentages are of total Austrian GHG emissions. CH_4 and N_2O are unwanted side products of combustion, created in small amounts only. Transport emissions consist almost exclusively of CO_2; the share of N_2O is only 1.2% and of CH_4 less than 0.1% of transport-related GHG emissions.

In 2010 GHG emissions caused by industrial processes were ranked second at 13% (11 Mt CO_2-eq.) after the energy sector. Emissions attributed to this sector include only process-related emissions (industrial processes during which GHGs are emitted); energy-related emissions are attributed to energy conversion (fossil fuel use, see paragraph above). Process emissions are divided up as follows: at 5.5 Mt CO_2 (2010) iron and steel production accounted for approximately 6.5% of Austrian GHG emissions. Ammonia production from natural gas accounted for 540 kt CO_2. Emissions of N_2O, a by-product of ammonia oxidation during nitric acid production, were reduced to 64 kt CO_2-eq., as the only Austrian plant was fitted with devices for the catalytic reduction of the emerging N_2O. In cement production, heating of carbonate rock releases CO_2, which accounted for 1.6 Mt CO_2 or almost 2% of total Austrian GHG emissions in 2010. Limestone production accounts for 574 kt CO_2. Magnesium sintering and "limestone and dolomite use" each account for approximately 300 kt CO_2, the latter being additives in blast furnace processes. The emissions of **fluorinated gases** (fluorinated hydrocarbons, F-gases) also relate primarily to industrial processes. With atmospheric lifetimes of several hundred years, F-gases have a strong climate effect. The refrigeration sector, including stationary and mobile cooling appliances, air conditioning units, and heat pumps, has seen the largest growth in F-gases. Under the terms of a European directive, only F-gases with GWP under 150 are permitted to be used in new appliances as of 2011. The use of F-gases in other areas (excluding extinguishing agents and in electrical switching stations) is decreasing, although older appliances and remaining stock are still causing emissions.

In **agriculture**, significant emissions of CH_4 and N_2O occur from enteric fermentation, manure management, and soil (emissions from energy use are attributed to the energy sector). In 2010 agriculture was responsible for 7.5 Mt CO_2-eq. or 8.8% of Austrian GHG emissions. The most significant sources of agricultural GHG emissions in 2010 were enteric fermentation from cattle (CH_4 emissions contributed 3.9% to total Austrian GHG emissions) and N_2O emissions from soil cultivation (3.4% in 2010). Manure management is responsible for both CH_4 and N_2O emissions (0.4% and 1%

of the Austrian total, respectively). Although forests tend to cause lower N_2O emissions than agricultural land, their contribution to Austrian emissions is still relevant due to the large forested area in Austria. The calculation of N_2O budgets from the landscape scale to the continental level is an unresolved challenge.

Biomass, in particular wood in forests, is a significant carbon repository. In Austria, this repository has tended to grow, and **forest-biomass has in the past, in most years, constituted a significant CO_2 sink**; however, in the last few years carbon sequestration has been on the decline and, in some years, has come to a complete halt. Austria has almost 4 million ha of forest (47.6% of its territory); thus, a large carbon stock (1990: 1 243 ±154 Mt CO_2 or 339 ±42 Mt C in biomass and 1 698 ±678 Mt CO_2 or 463 ±185 Mt C in soil), which is preserved due to sustainable forest management. Since the 1960s forest area has been increasing at all altitudes, particularly at altitudes of more than 1 800 m above sea level. As a result of climate change (increasing length of the vegetation period), improved nutrient availability (atmospheric nitrogen input) and an optimization of forest management, wood stocks are currently at record levels (2007/2009: 1 135 million cubic meters). However, due to increased felling of trees and the removal of particularly fast growing stocks, average productivity is on the decline.

Because of the release of landfill gases (CH_4 and CO_2, but also CFCs and N_2O) the **waste management** sector also causes a non-negligible share of GHG emissions. GHGs are emitted by waste incinerators and sewage treatment. CH_4 emissions result from anaerobic conversion processes of biologically degradable carbon compounds; their avoidances are an urgent priority for sustainable climate protection in waste management.

Modelling emissions from residual waste treatment for 2006 resulted in 1 250 kt CO_2-eq., or 1.5% of total Austrian CO_2 and CH_4 emissions (84 220 kt CO_2-eq.). Compared to 1990 levels, sectoral emissions have been decreasing due to emission reductions in landfills from originally 2 030 kt CO_2-eq., which indicates a decrease of more than 38%. As a result, the sector-specific emissions have decreased by approximately 18% to 0.89 Mg CO_2-eq./ton of residual waste.

The increase in total emissions since 1990 can be explained by the emissions of a few sectors. Major increases were observed in the transport sector, some of which can be explained by fuel exports "in the tank" (fuel tourism). Due to lower fuel prices in Austria, trucks in transit (and also passenger cars) tend to purchase considerable amounts of fuel in Austria (which are then apportioned to Austrian emissions) even

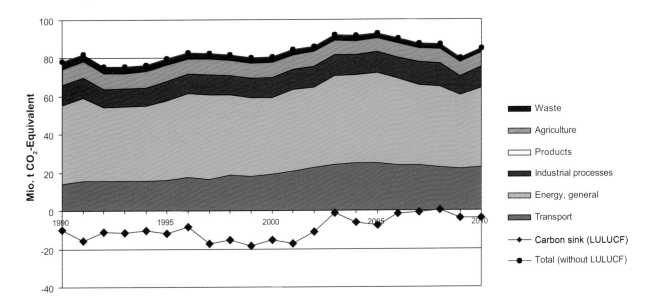

Figure S.1.4. Officially reported greenhouse gas emissions in Austria (according to the IPCC source sectors with especially defined emissions for the Transport sector). The brown line that is mainly below the zero line represents carbon sinks. The sector "Land use and land use change" (LULUCF) represents a sink for carbon and is therefore depicted below the zero line. In recent years, this sink was significantly smaller and no longer present in some years. This was mainly a result of higher felling; and changes to the survey methods contributed to this as well. Source: Anderl et al. (2012)

though much of the resulting distances are covered outside the country. It has been estimated that these fuel exports account for up to 30 % of transport-related CO_2 emissions, although these estimates are subject to high levels of uncertainty. Since the 1990s fuel prices in Austria have been consistently lower than in major neighboring countries. Conversely, carbon sinks have been lost: forests, which were active carbon sinks in the 1990s, lost effectiveness around 2003, when forests stopped accumulating CO_2 due to improved use of biomass (Figure S.1.4).

Since 1999 the **concentration of atmospheric CO_2** and since 2012 the **concentration of CH_4** have been measured continuously at the Hoher Sonnblick observatory (3106 m above sea level) within the framework of WMO's Global Atmosphere Watch-(GAW-) Program. In winter the concentration of CO_2 is higher than in summer due to higher emissions and lower levels of absorption by vegetation. Average annual values have been rising continuously, from 369 ppm (2001) to 388 ppm (2009) (Figure S.1.5). Data on the **ozone column** have been available at Sonnblick since 1994. Values are comparable to those measured in Arosa, Switzerland (±4 Dobson Units). Both location sites exhibit high interannual fluctuations, attributable to meteorological factors.

Inventories of particulate matter (PM) releases have been developed in view of the negative health effects of PM. Yet, in combination with knowledge of the chemical and physical characteristics of the emitted particles some conclusions can

be drawn regarding their climate relevance. The Austrian PM inventory assesses emissions of primary aerosols, that is, direct particle emissions in the atmosphere, but not of particles that develop from gaseous substances via atmospheric reactions and condensation of gases on particles.

Transport emissions, which account for approximately 44 % of PM2.5[5] emissions, include combustion products primarily from diesel motors (mainly diesel exhaust particulates) and, to a lesser extent, suspended particles from road dust.

Emissions from small heating installations (approximately 30 % of PM2.5 emissions) mainly include emissions from heating systems that use solid fuels, particularly wood (coal is hardly used as a fuel any more). Old heating systems and single stoves in particular cause significant emissions; and, due to their long lifespans, these appliances will continue to play a role in emissions for quite some time to come. With regard to emissions from **domestic heating**, elementary carbon (EC; soot), a part of the aerosol that has a particularly strong climate effect, is a significant component of PM2.5. The emissions, for example, from a typical Austrian tiled stove for various types of wood and wood briquettes have a soot component of 9.8 % (larch logs) and 31 % EC (soft wood briquettes). The emission factors of different biomass combustion systems were de-

---

[5] PM2.5 are particles with an aerodynamic diameter of less than 2.5 micrometers

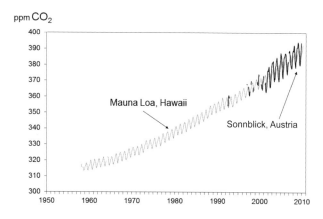

Figure S.1.5. Time series of CO_2 at Sonnblick Observatory (black line) in comparison with the measurements at Mauna Loa Observatory (grey line) for the last 50 years. Source: Böhm et al. (2011)

termined in the course of a bench test under various realistic operating conditions. Modern biomass-based heating systems have very low emissions; older single stoves and log burners, many of which are still in use, have high soot emissions. As soot is able to absorb radiation and thus have climate impacts, the advantage of avoiding fossil CO_2 emissions by using biomass fuels is diminished.

Particle formation (nucleation) in the atmosphere is an important parameter for the climate relevance of aerosols. **Secondary inorganic aerosols** (mainly sulphates, nitrates) and **secondary organic aerosols** (SOA) are formed in the atmosphere mainly through photochemical reactions on the part of precursor gases (e.g., NH_3, NO_x, SO_2, volatile organic compounds, VOC). Currently, no estimate of the annual contribution of secondary aerosols to the amount of aerosols in Austria is available. SOA are particularly important and are currently the subject of intense scientific scrutiny. Due to the long-distance transport of precursor gases, there can be very high background concentrations of ozone (O_3) and aerosol particles.

The data on aerosol mass concentrations collected by monitoring networks are alone insufficient to draw conclusions about the climate relevance of aerosols; however, together with other parameters (typical size distributions, meteorological conditions, chemical composition), they can be used to estimate climate-relevant aerosol characteristics. The atmospheric concentration of aerosols depends on emissions (see above), long-range transport, and meteorological and dispersion conditions.

The **chemical composition** of atmospheric aerosols which, via their refractive and hygroscopic properties, also influences their climate-relevant parameters contains information about

sources and chemical transformations in the atmosphere. With the help of a "macro-tracer" model, street dust and road salt, inorganic secondary aerosol, wood combustion, and traffic have been identified as the most important **aerosol sources** in Austria, although the relative contribution of the individual sources varies both regionally and temporally. The contribution of **wood smoke** to organic carbon (OC) in aerosol was between one-third and 70%, the contribution to PM10 from 7–23%. Under particular conditions "brown carbon" (BrC) from biomass fires can considerably exceed soot from traffic sources.

The Sonnblick observatory at 3 106 m above sea level is one of the most important background monitoring stations for aerosols and gases in Austria. Measurements of the chemical composition of aerosol demonstrate the changes that have taken place over the last 20 years and also the differences between the aerosol in the free troposphere (winter) and the planetary boundary layer (summer; Figure S.1.6). Long-distance transport of air masses (containing, for example, Saharan dust) can be observed all year round. The aerosol at Sonnblick was also investigated regarding aerosol-cloud interaction. The "**scavenging efficiency**" of soot (i.e., the fraction of atmospheric soot that can be found in droplets) is lower than that of sulphate (on average 54% compared with 78% on the Rax mountain at 1 680 m above sea level); yet a significant portion of the soot enters the cloud water by this process, where it can influence the radiative properties of clouds. Under conditions of 90% relative humidity, calculations of the **direct** effect of the Sonnblick aerosol resulted in **radiative forcing** of between +0.16 W/m² (assuming a ground covered by old snow) and +11.63 W/m² (fresh snow).

Carbonaceous aerosol has also been measured continuously at Sonnblick since 2005, and shows annual variations and concentrations similar to sulphate. Organic material (OM) accounts for the largest contribution to total carbon (TC). Approximately 10% of OM can be attributed to wood burning (summer: 4%; winter: 23%).

Because of the indirect effect of aerosol on the radiation budget, knowledge about cloud formation processes and cloud condensation nuclei (CCN) are extremely important. In Austria, CCN have been measured in various places (e.g., Rax, Sonnblick, Vienna). Long-term measurements of CCN in Vienna show that concentrations (at 0.5% supersaturation) are between 160 cm³ and 3 600 cm³, with an average of 820 cm³. Although seasonal variations have not been observed, CCN concentrations demonstrate large short-term variations, which result from different meteorological situations (stable weather conditions, passage of weather fronts).

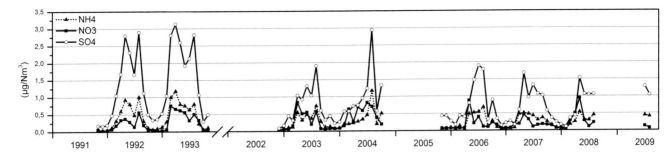

Figure S.1.6. Temporal variation of monthly mean values of particulate sulphate, nitrate and ammonium at the Sonnblick Observatory from 1991 to 2009. Sources: Kasper and Puxbaum (1998); Sanchez-Ochoa and Kapser-Giebl (2005); Effenberger et al. (2008)

Overall, due to their complex processes and interactions, the influences of aerosols on climate are a considerable scientific challenge. They are the largest uncertainty factor in determining radiative forcing.

S.1.3   Past Climate Change

To enable current climate change to be seen in context, reference is made to natural climate changes that have significantly shaped the current geological period, the Quaternary. When interpreting these climate developments, consideration must be given to the fact that the relevance of climate changes for humans depends fundamentally on population and lifestyle. For example, during the Pleistocene humans had not yet settled and population was at about 1 % of current levels.

The Pleistocene, which began 2.6 million years ago and ended 11 700 years ago, was shaped by an interplay between long glacial periods and short interglacial periods, controlled by changes in the parameters of the earth's orbit (shape of its orbit, and tilt and orientation of its rotation axis). The glacial periods were characterized by a climate of enormous variability, much larger than the climate variations that took place during the Holocene. The Dansgaard-Oeschger events (changes between very cold stadials and comparatively warm interstadials) known from Greenland ice cores, originated in instabilities in the large ice sheets and their interaction with deep water circulation in the Atlantic. This underlines the synchronous nature of high frequency glacial climate change at the supra-regional level. During the coldest phases of the glacial periods (the stadials) the Alpine foothills experienced arctic climate conditions with very cold winters. The warm periods were accompanied by abrupt decreases in seasonality (milder winters), but were initially, around 75 000 years ago, too weak to permit extensive reforestation in Austria. Toward the end of the last glacial period (Würm), approximately 30 000 years ago, a glacier advance beyond the Alpine foothills began. To

date, reliable paleo-climate data for the Alps do not exist: however, it is assumed that the average annual temperature was at least 10 °C below the corresponding temperature during the Holocene, combined with a significant decrease in precipitation toward the east.

Some 19 000 years ago the glaciers in the Alpine foothills and large Alpine valleys rapidly disintegrated. A number of regional and local glacial advances, primarily in the large tributary valleys, occurred in line with the climate developments in the North Atlantic–European region. Approximately 16 500 years ago, precipitation in the central Alpine region was reduced to between half and one-third of current values and the summer temperature was around 10 °C below current values. At the snow line of the glaciers that were then in existence, the ablation period lasted only approximately 50 days, around half the current duration. Winters were very cold and dry and comparable to present-day winters in the Canadian Arctic. Around 14 700 years ago a period with considerably more favorable inter-stadial conditions began within only a few decades, during which time, forests returned to north Alpine valleys and foothills. About 12 900 years ago the massive climate setback of the Younger Dryas began, the last significant cold phase in the northern hemisphere, which ended within a period of a few decades, 11 700 years ago. In the Alps, this period was characterized by considerable glacial advances in upper valley areas, a significant lowering of the timber line, and increased geo-morphological activity due to permafrost in non-glaciated areas. The snow line was 300–500 m lower and the lower limit of permafrost at least 600 m lower than during the middle of the 20th century. Summer temperature was approximately 3.5 °C lower than during the middle of the 20th century, and annual temperature was lower still. In the central Alps precipitation levels were approximately 20–30 % lower than today, whereas the outer regions of the Alps may have been wetter.

The climate during the Holocene. The first centuries of the Holocene were characterized by glacial advances that were

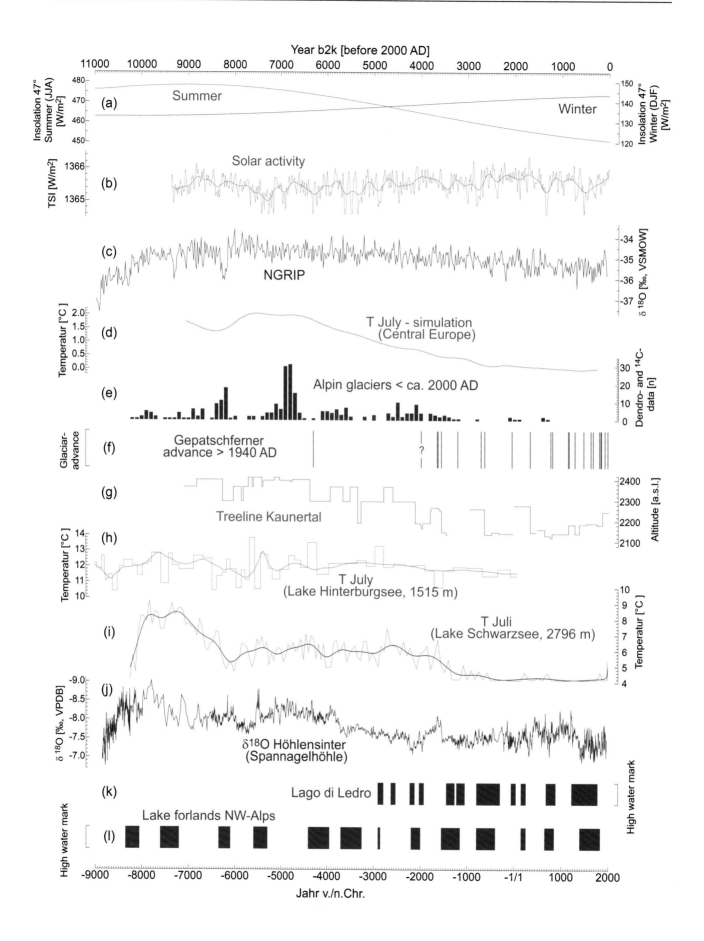

considerably more extensive than during the "little ice age": there may have been permafrost in regions up to 200 m lower than today. Summer temperatures during the earliest Holocene were 1.5–2 °C lower than in the 20th century, whereas precipitation levels were comparable to today's. A selection of reconstructions of climate parameters from proxy data can be found in Figure S.1.7.

The cold start to the Holocene was followed by significant warming. According to reconstructions from various Austrian climate archives, temperature values during the first two-thirds of the Holocene were above the 20th century average. All proxy data show a long-term temperature decrease of around 2 °C from the early to middle Holocene maximum values (i. e., until approximately 7 000 years ago) to the preindustrial period. The undisputed cause of this cooling trend is the decrease in solar radiation to the northern hemisphere in summer, caused by orbital variability. In contrast, another much discussed climate forcer, solar activity, demonstrates no comparable long-term trend. Analysis of precipitation during the Holocene shows no long-term development to date; rather multi-decadal and multi-century periods with higher and lower levels of precipitation alternating. Periods with increased precipitation coincided with phases of reduced solar activity.

During the last 11 000 years, glaciers in the Alps were characterized by long periods with comparatively small expansion during the early and middle Holocene (up to around 4 000 years ago) and multiple and extensive advances in the following millennia, which cumulated in the large glacier extent during the "little ice age" (from approximately 1260 to 1840 AD). The extent of glaciation during the early and middle Holocene was beneath and above current levels, several times over. However, alpine glaciers are currently not in balance with the climate they are controlled by, which is manifest in the strong melting that has been observed. It is thus difficult to directly compare the current glaciation extent with earlier ones with regard to climatic boundary conditions.

The climate of the last two thousand years. During the last 2 000 years there was a succession of warmer and colder periods, which, on average, were colder than during the beginning and middle of the Holocene. This time can be roughly

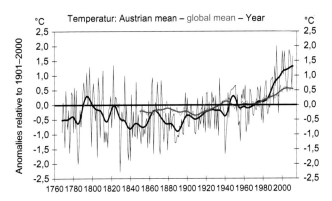

Figure S.1.8. Anomalies in the annual mean air temperature for Austria (1768 to 2011) and the global mean temperature relative to the respective 20th century mean (1850 to 2011). Single values and smoothed values using a 20-year Gaussian low pass filter. Source: Böhm, (2012), source HISTALP (http://www.zamg.ac.at/histalp) and CRU-data (http://www.cru.uea.ac.uk/data)/)

divided into four periods, starting with the relatively stable and mild Roman warm period (from approximately 250 BC to 300 AD). This was followed by an unstable period, through to the end of the Roman era and during the early Middle Ages (from approximately 300 to 840 AD), characterized by moist and cold summers. This, in turn, was followed by a warmer, more stable period (Medieval Warm Period, from approximately 840 to 1260 AD). Between 1260 and 1860 AD it was considerably colder: only during individual decades were proven warmer temperatures demonstrated. Because of the generally large glaciation shown to have existed during this period, it is also referred to as the "little ice age." Several minima of solar activity and also climate-effective volcanic eruptions occurred during this period. The significant increase in temperatures during the 20th century measured by instrumentation is reflected by the natural climate archives, even though many proxy datasets ended in around 2000 AD and do not include current climate developments in their entirety.

The instrumental period. The Austrian network of meteorological monitoring stations is such that long-term climate change in the 19th and 20th centuries can be accurately described. The oldest evaluable measurement series in Austria, the Kremsmünster series, dates back as far as 1767 and is one

Figure S.1.7. (Left page) Holocene environmental records and proxy-based climate reconstructions from Austria, the Alps and Greenland in comparison with selected climate forcings. a) evolution of insolation during summer (June-July-August) and winter (December-January-February) at 47°N; b) reconstruction of solar variability for the last 9 000 years; c) oxygen-isotope record of the NGRIP ice-core, central Greenland; d) simulation of the temperature evolution in July in central Europe over the last 9 000 years until the pre-industrial period ; d) dendrochronologically, i. e. calendar-dated, and 14C-dated evidences for shorter glaciers than today (≈1990/2010 AD); f) established advances of the glacier Gepatschferner beyond the glacier's size in 1940 AD; g) tree-line record in the Kauner valley based on wood remain findings; h) chironomid-based reconstruction of July temperature from lake Hinterburg, Switzerland; i) chironomid based reconstruction of July temperature from Schwarzsee ob Sölden; j) oxygen-isotope record of speleothems from the Spannagel cave; k) lake-level high-stands of Lago di Ledro during the last 5 000 years; (l) lake-level high-stands in the foreland of the NW-Alps and the Jura. Source: Compiled for AAR14

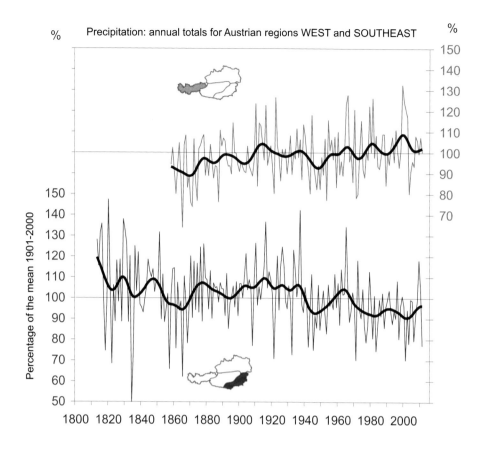

Figure S.1.9. Anomalies of the annual precipitation totals relative to mean of the 20th century for two Austrian subregions (top: "West", bottom: "Southeast"). Single values and 20-year smoothed values (Gaussian low pass filter). Time-series date back to 1813, but with differing starting dates, and continue through to 2011. Copyright by R. Böhm (2012), source HISTALP http://www.zamg.ac.at/histalp

of the longest continuous weather records in Europe. Data from stations in Vienna (old university observatory) and Innsbruck (university) can be used for climate analysis of the latter part of the 18th century. Of particular note is the high-alpine Sonnblick observatory, the weather records of which date back to 1886. The measuring station is situated at the peak of "Hoher Sonnblick," 3106 m above sea level, directly on the main ridge of the Alps.

In Austria, the temperature has risen by nearly 2 °C in the period since 1880, compared with a global increase of 0.85 °C. The increase can be observed particularly in the period after 1980, during which global temperatures rose by approximately 0.5 °C, compared with an increase of approximately 1 °C in Austria (virtually certain, Figure S.1.8; Volume 1, Chapter 3). Seasonal temperature developments did not always run parallel to the annual average; nevertheless, warming has occurred during all seasons since the mid-19th century, with lowest increases taking place in the autumn. The cooling effect of anthropogenic aerosols ("global dimming") likely played an important role in the temperature stagnating to decreasing during the three decades from about 1950 to 1980; this masked the effect of GHG emissions that were already on the increase.

Temperature developments in higher air layers, derived from homogenized radiosonde measurements, are very similar at 3000 m above sea level to developments at high alpine stations. The higher warming trend observed in the Alpine region, when compared to the global average, gradually decreases to typical mid-latitude warming trends in higher layers. A significant temperature decrease can be observed in the stratosphere (13 to 50 km above sea level) above Austria, and also globally.

Air pressure at lowland stations demonstrates a very long-term increase from the middle of the 19th to the end of the 20th centuries, which was replaced by an abrupt change in trend to decreasing air pressure in around 1990. Air pressure at Alpine high altitude stations is also influenced by the temperature of air masses below the measurement stations. Because of the warming of air masses below, these stations show a stronger positive pressure trend and no pressure decrease since 1990. This deviating trend in air pressure at high Alpine observatories, when compared with lowland trends, is a confirmation of warming that does not depend on thermometer measurements.

In the past 130 years the annual duration of sunshine at mountain stations in the Alps has increased by around 20 %, or more than 300 hours. The increase was higher in summer than in winter (virtually certain, Volume 1, Chapter 3). Due to increased cloudiness and an increase in air pollution from 1950 to 1980, the duration of sunshine in summer decreased significantly, particularly in valleys. The sustained trend of more sunshine since 1980 is accompanied by more and longer summer fair-weather periods.

Unlike temperature developments, **developments in precipitation demonstrate considerable regional differences during the past 150 years**: while levels of precipitation have increased by around 10–15 % in western Austria, the southeast has seen a decrease of similar proportions (Figure S.1.9). In inner Alpine regions and in the north, decadal variations predominate. All parts of Austria were particularly dry during the 1860s. Such values have only been reached or undercut in the southeast during the dry 1940s and the persistently dry decades after 1970.

There were decades with high levels of precipitation in the first half of the 19th century. This played an important role in the glacial advances during this period, which led to the two maximum glacier levels in around 1820 and in the 1850s. There was also high annual precipitation in the decades between 1900 and 1940 (almost continuous in inner Alpine regions and in the southeast, reduced in the west and interrupted by an arid phase in around 1930 in the north). This was followed by lower levels of precipitation in the north and in inner Alpine regions, that were followed again by a marked change in the 1970s and – particularly in northern and northeastern Austria – a new precipitation maximum in the first decade of the 21st century. Current levels of precipitation in the west are also at their highest since measurements began in 1858. In inner Alpine regions current precipitation levels are around the long-term 20th century average, in the southeast – in the course of the decreasing long-term trend – around 10 % below the long-term 20th century average.

In mountainous Austria, climate change at higher elevations is of great importance. The climate series from Hoher Sonnblick (3 106 m) are considered to be representative for a high mountain climate. According to these measurements, temperature increase in high mountain regions is similar to that in valleys, although a considerably stronger increase in sunshine can be observed, which can be attributed to European clean air measures. There has been a clear shift from snowfall to rain; at Hoher Sonnblick about 30 % of precipitation is presently rain. Average air pressure is increasing in the mountains – a sign that the lower-lying air masses are warming. A significant decrease in glacier volume and extent and melting of permafrost have also been documented (Volume 1, Chapter 5).

Austria has very good long-term meteorological measurement series, offering a high potential for the integration of data analysis and model simulations. International coopera-

tion would lend itself very well to the compilation of high-resolution datasets, both temporally and spatially, for the Alpine region and Europe. It would also be beneficial to strengthen less well developed measurement networks, such as those for measuring GHGs, aerosols, and radiation.

S.1.4   Future Climate Change

To make geographically detailed statements about the future climate, primarily regional climate models are applied, which are integrated into the results of global climate models. As with global models, the results of different models are analyzed to differentiate between robust and less robust results. There are a number of simulations of both past and future climate covering the Alpine region and Austria. In the following, simulations based on the A1B emission scenario – a scenario with a medium to large increase in GHG concentrations – will be the main focus of analysis. Using one scenario enhances the comparability of results. Using this particular scenario makes the changes more apparent than a more optimistic scenario (with lower emission increases) on the one hand, and is closer to current emissions trends, on the other. Furthermore, this choice is more consistent with the precautionary principle, described earlier.

In Austria a further temperature increase is to be expected (very likely, Volume 1, Chapter 4; see also Figure S.1.10). In the first half of the 21st century temperature will increase by approximately 1.4 °C compared to today's temperature level. This increase is not greatly affected by assumptions about future greenhouse gas emissions because of inertia in the climate system and the longevity of greenhouse gases in the atmosphere. The temperature development thereafter, however, is strongly dependent on anthropogenic greenhouse gas emissions in the coming years, and can therefore be influenced (very likely, see Volume 1, Chapter 4). Figure S.1.10 shows the temperature development in Austria from 1800 to 2100 as a deviation from the average temperature during the period 1971 to 2000, for the A1B emissions scenario. The anticipated medium temperature increase in the Alpine region for the period 2021 to 2050 compared to the reference period 1961 to 1990 is +1.6 °C (0.27 °C per decade) in winter and 1.7 °C (0.28 °C per decade) in summer. Further and accelerated warming of the Alpine region is projected by the end of the 21st century in the A1B emissions scenario.

In the 21st century, an increase in precipitation in the winter months (around 10 %) **and a decrease in the summer months** (around 10–20 %) **is to be expected** (likely, Volume 1, Chapter 4). The annual average shows no clear trend, as the

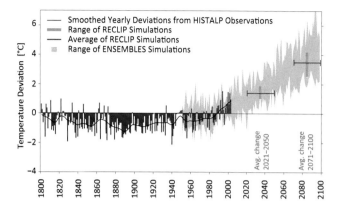

Figure S.1.10. Mean surface temperature in Austria since 1800 (instrumental observations, in colour) and expected temperature development until 2100 (grey) for one of the higher emission IPCC scenarios (IPCC SRES A1B), shown as a deviation from the mean 1971 to 2000. Columns represent annual means, the line smoothed values over a 20 year filter. The slight temperature drop until almost 1900 and the strong temperature increase (about 1 °C) since the 1980´s can be clearly seen. For this scenario, a temperature increase of 3.5 °C until the end of the century is expected (RECLIP Simulations). Source: ZAMG

Alpine region lies in a transition region between two zones with opposing trends (likely, Volume 1, Chapter 4). Figure S.1.11 shows how precipitation develops in Austria (divided into two regions, northwest and southeast) for winter and summer from 1800 to 2100, as a deviation from the average during the period 1971 to 2000. On the basis of several models, a tendency toward precipitation increase north of the Alps in spring, summer, and autumn can be expected, whereas the southern and western parts of the Alpine region exhibit decreases. However, these geographically differentiated precipitation changes are subject to a high level of uncertainty. Figure S.1.12 shows the annual cycle of changes for the periods 2021 to 2050 and 2069 to 2098 based on an ensemble of models. Although the trend described toward increased precipitation in winter and decreased precipitation in summer can be identified in the median in the first half of the century, models show no agreement about the direction of the changes in this period (left panel). Toward the end of the 21st century, the A1B scenario shows a clear trend toward drier conditions in summer (approximately 20 % less precipitation) and wetter conditions in winter (approximately +10 %).

Similar to the trends in precipitation, global radiation (shortwave solar and sky radiation) shows almost no change throughout the year until the middle of the 21st century. However, toward the end of the 21st century a significant increase can be observed in summer and a decrease in winter (Figure S.1.12). This is consistent with projections of precipitation, as precipitation-producing clouds shield solar radiation.

The clear decrease in relative moisture, approximately 5 % by the end of the century, is a result of the lower amounts of precipitation during summer months. Projections of wind speeds are subject to high levels of uncertainty – models project both positive and negative trends – although most models anticipate a decrease in wind speeds rather than an increase by the end of the century (Figure S.1.12).

Methodological advances to optimize the interface between purely physical climate modelling and the ever more important investigation of regional impacts of climate change are necessary and promise a comparatively quick increase in quality of climate impact research. This also requires a better understanding of small-scale processes and extreme events.

S.1.5   Extreme events

Extreme weather events can have significant impacts on nature, infrastructure, and human life. They are, however, statistically difficult to determine, as changes in rare events can only be identified in long time series – the more extreme the event, the longer the time series required. Uncertainty regarding frequency and intensity of small-scale extreme events such as thunder- or hailstorms also results from a lack of geographical and temporal resolution of the available climate data and models. In Austria statistical analysis of extreme events is rendered difficult by the fact that most of the older time series with daily data were lost in World War II and only time series with monthly mean values remain. Furthermore, the strong non-linearity of the phenomena that lead to extreme events has still not been fully resolved scientifically and remains a challenge. However, some statements can be made about extreme events, particularly if the considerations or calculations are based on the atmospheric processes that underlie such events.

Temperature extremes are increasing (heat). Analyses based on homogenized daily temperature extremes since 1950 show an increase in hot days and warm nights across Austria. In parallel to these developments, cold days and cold nights have decreased significantly. With the increase in temperature extremes, the number of frost days and ice days has decreased. **In the 21st century temperature extremes, for example the number of hot days, will increase significantly** (very likely, Volume 1, Chapter 4). According to model projections, temperature in Austria will increase by 4 °C during hot periods in summer by the end of the 21st century. At the same time the frequency of heat waves will increase from around 5 to around 15 per year by the end of the century. At the two hottest Vien-

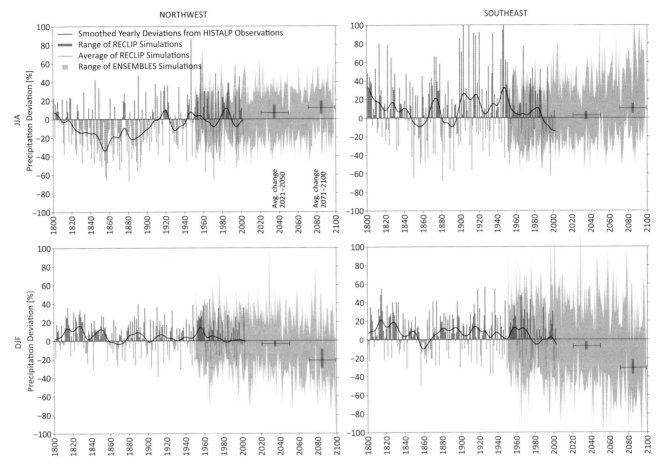

Figure S.1.11. Precipitation development in Austria since 1800 (instrumental observations) and expected development to 2100, shown as a deviation from the mean 1971 to 2000. Bars at the top show the winter season (December to February, DJF), the bars at the bottom the summer season (June to August, JJA). The region of Austria is divided (north-west and south-east) into two regions. The observational data for the past stem from the HISTALP database, scenarios for the future from the 22 ensemble simulations (www.ensembles-eu.org, grey bars for single years) and from reclip: century (http://reclip.ait.ac.at/reclip_century, coloured bars for the time slices 2021 to 2050 and 2071 to 2100)

nese stations, the number of hot days will increase from on average around 15 currently to around 30 by the middle of the century and between 45 and 50 by the end of the century. At the same time, the number of cold nights with frost in the inner city will decrease from around 50 events currently, to fewer than 40 by the middle of the century and just over 20 by the end of the century (Volume 1, Chapter 2; Volume 1, Chapter 4).

Cities in particular will be affected by temperature extremes, as the effects of urban heat islands are superimposed on to climate change. In Vienna, as an example of **urban space**, a statistically significant increasing trend in the temperature difference between the city and its surroundings has been observed since 1951. High temperatures during the day and less cooling during the night lead to negative health effects in the urban population (Volume 1 Chapter 5; Volume 3, Chapter 4). In future, with further increases in temperature, heat stress will

present a significant challenge for urban areas. In this connection an increased demand for energy for cooling can be expected, while at the same time the demand for heating will decrease (Volume 3, Chapter 5). Urban-planning measures, such as compact building structures with ample ventilation, adequate shade, greening roofs, facades, and streets and light-colored surfaces, can significantly reduce urban heat stress. Given the long-term nature of urban planning and the expected increase in heat stress for the population, timely planning of such measures is of utmost importance (Volume 1, Chapter 5).

Due to insufficient data, statements about the change in frequency of damage-inflicting precipitation events are subject to significant uncertainty. **Extreme-value indices**, derived from homogenized time series of daily precipitation sums, and intensity of precipitation or maximum daily precipitation sums, show neither statistically significant nor consistent trends to

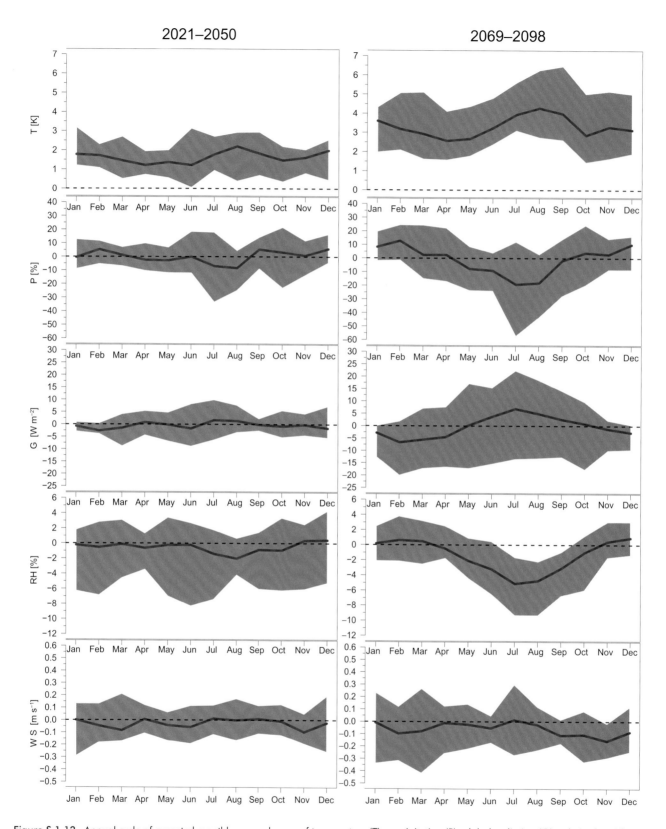

Figure S.1.12. Annual cycle of expected monthly mean change of temperature (T), precipitation (P), global radiation (G), relative humidity (RH), and wind speed (WS) in the Alpine region relative to the reference period 1961 to 1990 for the SRES A1B-Scenario. Left column: 2021 to 2050, right column: 2069 to 2098. The blue line indicates the median, the grey shading the 10–90th percentile range of the multi model ensemble. Source: Gobiet et al. (2014)

date. Large-scale extreme precipitation events have tended to increase since 1980.

Climate models indicate more extreme events in future. However, to date almost all model studies on future precipitation extremes consider only changes in mean values on a seasonal basis or the probability of exceeding defined percentiles for large areas. Statements regarding the intensity and frequency of future extreme events become more robust the larger the spatial or temporal scale of the extreme event (e.g., large-scale dry spells; Volume 1, Chapter 4). In general, the more detailed analyses of precipitation extremes are, then the larger the uncertainties and differences between models. Results of simulations of high-resolution regional models often demonstrate geographical patterns of climate signals for the future that are of such complexity that a clear interpretation is impossible. This is particularly true for extreme convective precipitation events, as frequently occur during fair weather in summer or in the Alpine foothills (Volume 1, Chapter 3; Volume 1, Chapter 4).

The potential for increased probability of **heavy precipitation** can be deduced from a warmer future atmosphere containing more water vapor. From autumn to spring extreme precipitation events will probably increase (Volume 1, Chapter 4). For Central Europe, models show that the number of days with precipitation and the intensity of precipitation will increase by 10% during the winter. There will, however, be differences across Central Europe as to whether multi-day heavy precipitation events, which pose a considerable risk of floods due to soil water saturation, will increase or decrease. During summer months in Austria, an increase in intensity of 17–26% of 30-year precipitation events has been projected for the period 2007–2051, in comparison with the period 1963–2006. The increase in precipitation intensity during autumn appears to be particularly distinct in the southeast and east of Austria – this may be an indication for a change in frequency of weather patterns in the eastern Alpine region (Volume 1, Chapter 4).

The climate of the Mediterranean is of particular importance for flood risk in Austria because air masses can quickly become enriched with moisture over the Mediterranean Sea and carried into the Alpine region. In particular, the pronounced precipitation maximum in October in southern Austria can be attributed to low pressure areas moving in from the Mediterranean (particularly those on "Vb-tracks") and the high surface temperature of the Mediterranean Sea at that time of year. Many disastrous floods in the past have been attributed to cyclones on Vb-like tracks, including the events in July 1997, August 2002, and August 2005. Although it is

not possible to quantify the potential future changes in the frequency of precipitation-intense Vb-track cyclones, it is clear that a warmer Mediterranean Sea in future could lead to more precipitation-intense Vb-cyclones and consequently increase the risk of extreme floods in Austria (Volume 1, Chapter 4).

No long-term increase in storm activity, deduced from homogenized daily air pressure data, **could be detected**, despite a number of major storm events in the past years. For the future, no change can yet be inferred. Models indicate a weak decrease in maximum daily wind speeds of 20-year events. However, the details of the results are uncertain and, depending on the model, range from +10% to –10% (Volume 1, Chapter 4).

Changes in the frequency or intensity of thunderstorms and hail are among the most relevant but also most challenging questions in climate research. Analysis of weather conditions conducive to hail storms shows a weak but statistically significant increase in the potential for hail storms in Central Europe over the past decades. Regional models do not indicate any change for the future (2010 to 2050) in this regard (Volume 1, Chapter 4).

Investigations into aridity show a three-fold increase in the likelihood of the occurrence of drought in future climate, 2071–2100, compared to the past (1961–1990) for the SRES A1B scenario. The length of dry periods also increases and soil moisture content drops below present levels. As models cannot yet determine, with sufficient reliability, regional precipitation, local soil moisture, and the persistence of atmospheric circulation, these projections remain subject to significant uncertainty (Volume 1, Chapter 4).

Lake Neusiedl, a shallow lake with highly variable water levels depending mainly on precipitation (Steppensee), will be particularly affected by aridity. Lake Neusiedl has significant influence on regional climate, is important for tourism, water sports, shipping, and fisheries and has unique fauna and flora. Despite infrastructural measures being in place to regulate water levels, the water budget is influenced mainly by natural factors that depend largely on climate. Slightly lower precipitation levels during the period 1997–2004 with increasing temperatures led to continuously decreasing water levels in Lake Neusiedl. In particular, the low water level in the year 2003, caused by extremely low annual precipitation and high air and water temperature, raised the question as to whether, subject to future climate conditions, the lake would dry up. Studies showed that warming of 2.5°C would lead to an increase in evaporation of over 20%. To compensate for this loss of water, precipitation would need to increase by approximately 20%, which, according to current climate scenarios, is unlikely. A succession of dry years in future could lead to very low water levels or even to the lake drying up. Water management measures could temper but ultimately not hinder this process. It is expected that even a moderate decline in the water level would have significant ecological and economic impacts. Therefore, both mitigation (supplying additional water) and adaptation strategies, such as diversifying tourism activities and extending the spring and autumn seasons, are being considered (Volume 1, Chapter 5).

S.1.6 Thinking Ahead: Surprises, Abrupt Changes and Tipping Points in the Climate System

Unexpected weather situations and new surprising research results often help to close knowledge gaps. A surprising development in recent times was the hypothesis that the decline in sea ice in the Arctic might directly influence the duration of winter in Europe, the snow pack, and the temperature levels, and, in particular, that it could lead to more frequent intrusion of cold Arctic air masses causing extremely cold conditions in Europe. The decline in Arctic sea ice is an example of unexpected abrupt changes in the climate system that could also occur in other elements of the system. In particular, exceeding the so-called tipping points could lead to positive feedback loops and thus to irreversible and very extreme changes in the global climate system (Volume 1, Chapter 5).

Such disruptions are hard to predict; however, it is known that various components or phenomena of the climate system have experienced abrupt and partly irreversible changes in the past. The question regarding the occurrence of tipping points in the future can be neither definitely negated nor confirmed. However, it is assumed that increasing temperatures generally and warming of more than 2 °C over pre-industrial levels in particular, will increase the likelihood of abrupt changes. Tipping points can occur not only in the climate system, but also in other natural, political, economic, and social systems as a result of climate change. Such processes imply enormous impacts on human civilization, and the precautionary principle requires that they be considered when political, economic, and societal decisions are being made (Volume 1, Chapter 5).

S.2 Impacts on the Environment and Society

S.2.1 Introduction

Humans and the environment are connected inseparably. The impacts of climate change therefore need to be considered in an integrated manner for the human-environment system (Figure S.2.1; Volume 2, Chapter 1).

The current epoch is also referred to as the Anthropocene. There are (with a few exceptions) very few places and subsystems on the planet that are not influenced by human activity. As humans have become the dominant driver of change on our planet, the term Anthropocene (the Greek "anthropo" meaning human and "cene" meaning new) was coined for the current geological epoch (Volume 2, Chapter 1). The manifold human influences on the environment – of which anthropogenic climate change is just one aspect – make it difficult to link observed changes to changes in the climate system in some areas (Volume 2, Chapter 4). To understand the complexity of the current situation and identify possible solutions with regard to future developments, human being must be taken into account as the central driver at all scales (Volume 2, Chapter 1).

Climate change has both **direct and indirect impacts** on humans and the environment (Volume 2, Chapter 1). Direct impacts occur where changes in climatic parameters such as temperature or precipitation have immediate effects. Indirect impacts are impacts in which climate change becomes effective through its influence on another process in the system. In the case of the impacts of climate change on soils, for example, there needs to be differentiation between i) direct effects of temperature on soil-borne processes (such as weathering or the effects of increased extreme precipitation on soil erosion) and ii) indirect effects via the climate influence on vegetation rooted in the ground (where, for example, dead organic material influences humus formation) (Volume 2, Chapter 5). In certain cases the indirect effects of climate change can have stronger impacts than the direct effects (Volume 2, Chapter 5; Volume 2, Chapter 6).

The causes and impacts of climate change are often decoupled both temporally and geographically. Humans are both affected by climate change and drivers of it. The local actions of every individual affect the global energy balance of the atmosphere. Global climate change associated with these effects, has very different characteristics regionally and locally; it has multiple consequences that often occur with significant delay. The same principle applies to climate protection. The effects of individual contributions to climate protection are not im-

Hydrosphere
Bd. 2, Kap. 2; Bd. 3, Kap.2

Biosphere
Bd. 2, Kap. 3; Bd. 3, Kap. 2

Atmosphere
Bd. 1, Kap. 2-5

Geosphere
Bd. 2, Kap. 4

Pedosphere
Bd. 2, Kap. 5

regional human-environment-systems

Social Concerns
Bd. 2, Kap. 6.2

Economy
Bd. 2, Kap. 6.3

Tourism
Bd. 2, Kap. 6.4; Bd. 3, Kap.4

Health
Bd. 2, Kap. 6.1; Bd. 3, Kap.4

Infrastructure
Bd. 2, Kap. 6.7; Bd. 3, Kap. 3

Settlement
Bd. 2, Kap. 6.6; Bd. 3, Kap. 5

Natural Hazards
Bd. 2, Kap. 6.5

Figure S.2.1. Interfaces between global drivers system und local / regional human-environmental systems as a response systems between the natural spheres and the anthroposhere

mediately discernible either temporally or spatially. Regions which contribute over-proportionally to climate protection do not experience a comparably larger reduction in warming or otherwise reduced climate impacts. This dilemma of geographical and temporal decoupling of cause and effect is probably the most salient factor leading to the poor grasp of the seriousness of global climate change that currently exists. This, in turn, leads to the unfortunate lack of acceptance of necessary measures, both in mitigation and adaptation terms, to deal with climate change. The geographical-temporal decoupling of cause and effect also complicates the question of polluter and damaged parties / beneficiaries as well as the global responsibility for climate change. On a global level the societies most vulnerable to climate change are often not the main polluters, whereas the actual polluters profit from many of the advantages resulting from climate change. This raises questions of climate justice (Volume 2, Chapter 1).

The complex impacts of climate change on the human-environment system can be described in terms of **vulnerability, resilience, and capacity**. The complexity resulting from the decoupling of causes and impacts, together with the non-linear interactions across spatial and temporal scales, means that a systematic approach must be taken to the analysis of climate impacts. Vulnerability describes the extent to which an exposed system is susceptible to disruptions or stress; it also refers to how restricted that system is in its ability to cope with or overcome these challenges. As such, it is a measure of the sensitivity of the human-environment system to the negative effects of climate change at any given stage; it also describes its ability or lack of ability to overcome the consequences of climate change. Vulnerability is counteracted by resilience. Resilience expresses the ability of an individual, society, or system to cope

with or overcome an adverse influence. The idea of resilience was based originally on the concept of ecosystems' ability to withstand disruptions without changing in structure or collapsing. More recently, the concept of resilience has also been used with respect to social systems, for example, in the field of natural hazards and risks. Here, resilience refers to the ability of individuals or social groups to compensate for external stress factors and disruptions, resulting from ecological, social or political influences and to be able to plan in a future-oriented manner. The nature of vulnerability and resilience implies that they are mainly used to refer to potentially negative changes in the system, while possible positive developments leading potentially to an improved state of the system are neglected. Therefore, separate reference is often made to the capacity of a system to pick up and develop a specific impulse toward an improved state of the system, described as "absorptive capacity". Here the focus is on capacity building, which then can contribute to the adaptation to changing conditions in the sense of adaptive capacity (Volume 2, Chapter 1).

Adaptation to climate change is necessary to cushion or deter negative impacts and avoid ruptures in the system. Despite all the efforts to mitigate a further increase in the human-induced greenhouse effect, climate change in the 21st century is inevitable; only its scale is still undetermined. Adaptation is a guiding principle that is essential for survival and that can contribute to avoiding ruptures in, or a collapse of, the human-environment system. Adaptation activities are goal-orientated and aim either to reduce risks or to achieve positive developmental potential. Mitigation and adaptation to climate change (Volume 3, Chapter 1) are closely connected – the need for adaptation becomes greater, the less mitigation measures take effect. The ability of a system to adapt depends

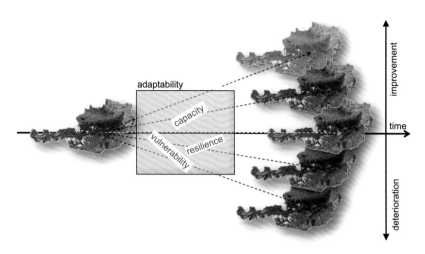

Figure S.2.2. Open concept of adaptability, based on the open risk concept. Possible future conditions that may exist in Austria are a function of its adaptability; Source: Coy and Stötter (2013)

on the one hand on vulnerability, resilience, and capacity, and on the other hand on the intensity of climate change (Figure S.2.2). In general, the adaptive capacity of a system needs to be considered in medium- to long-term time scales; it therefore possesses, comparable to the sustainability principle, a generation-spanning dimension (Volume 2, Chapter 1).

The concept of **ecosystem services** makes it possible to quantify some of the ecological impacts of climate change and their effects on society. The concept of ecosystem services – introduced in the Millennium Ecosystem Assessment – quantifies the services of ecosystems that are provided by nature and used by humans. Ecosystem services fall into four categories:

1. Provisioning services: products, which are directly removed from ecosystems (e.g. foodstuffs, drinking water, wood, combustibles, herbal medicines).
2. Regulating services: such as the regulation of climate and air quality, reduction of extreme events, and biological pest control.
3. Cultural services: such as recuperation, experience and education in nature, spiritual and aesthetic values.
4. Supporting services: services of ecosystems that are necessary to provide for the first three categories (e.g. photosynthesis, material cycles, and soil accumulation).

As ecosystems react sensitively to changes in climate and as the services that humans obtain from ecosystems are affected by these changes, ecosystem services provide a good indicator with which to evaluate the impacts of climate change on the human-environment system. Furthermore, (long-term) monitoring of ecosystem services provides the possibility of quantifying the indirect effects of climate change that are sometimes difficult to grasp (Volume 2, Chapter 1; Volume 2, Chapter 3).

S.2.2   Impacts on the Hydrological Cycle

Snow: The snow fall limit has retreated since 1980, and the change is particularly pronounced in summer. The retreat in winter is small compared to variability. These developments are in accordance with the considerably larger increase in air temperature in summer than in winter (Volume 1, Chapter 3; Volume 2, Chapter 2). Because of the increase in temperature, the snowline is projected to retreat by 300 to 600 m by the end of the century, or approximately 120 m per 1 °C of warming.

The duration of snow cover has decreased in the past decades, particularly at intermediate altitudes (around 1 000 m above sea level). As both the snow fall limit – and consequently the increase in snow pack and snow melt – are temperature-sensitive, a decrease in snow depth is expected at intermediate altitudes due to the continued increase in temperature (very likely, Volume 2, Chapter 2 and 4). Model calculations show an average decrease in snow cover duration of 30 days, for the 1 000–2 000 m altitudinal belt. At low levels (<1 000 m) and high levels (>2 000 m) the decrease will only be for approximately 15 days. Projections show that the south and southeast of Austria are particularly affected by the decline, with a future average duration of 70 days of snow cover. Snow cover comparable to today's levels will be found in areas shifted upwards by about 200 m by the middle of the 21st century (Volume 2, Chapter 2).

In lower and medium altitudes a climate-induced reduction in snow avalanches is expected. (Volume 2, Chapter 2 and 4). The decreasing amount of solid precipitation at lower-to-medium altitudes leads to reduced levels of fresh snow, which in turn reduces avalanche activity. At higher altitudes, levels of fresh snow could increase, although due to changing temperatures, a shift from powder snow avalanches to wet

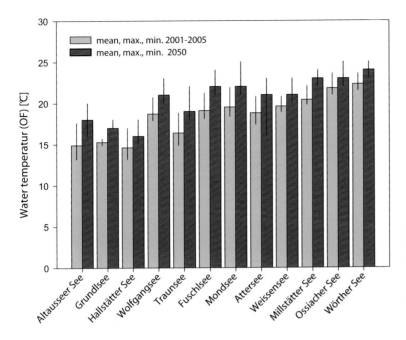

Figure S.2.3. Observed (green) and estimated (red) surface water temperatures (OF) in lakes for 2050 during the bathing season (June to September). The columns indicate the mean, the lines the maximum and minimum values between 2001 and 2005; the estimates for 2050 are based on a linear trend. Source: Dokulil (2009)

snow avalanches can be expected. When looking at changing avalanche activity, changes in forests must also be considered (Volume 2, Chapter 3). An increase in forest density at high altitudes can have a dampening effect on avalanche activity (Volume 2, Chapter 4).

Glaciers: All surveyed glaciers in Austria have significantly declined in area and volume in the period since 1980. In the southern Ötztaler Alps, the largest connected glacier area in Austria, for example, glacial area has shrunk from 144.2 km² in 1969 to 126.6 km² in 1997 to 116.1 km² in 2006 (Volume 2, Chapter 2). Between 1969 and 1998 Austrian glaciers lost a total of around 16.6 % of their area (Volume 2, Chapter 2).

Austrian glaciers have reacted particularly sensitively to summer temperature during the period of decline since 1980. It is therefore expected that **by the year 2030 the ice volume and the area of Austrian glaciers will have declined to half of the mean values of the 1985–2004**. In terms of future loss of glacier mass, the climate scenario chosen plays a quite minor role, as a substantial part of future loss of mass is a (delayed) consequence of past climate change. In the most favorable scenario, Austrian glaciers will stabilize at around 20 % of current ice volume by the end of the 21st century, whereas the extreme scenario leads to an almost entire melting of glaciers in Austria (Volume 2, Chapter 2). With rock glaciers, an increase in temperature at first leads to an acceleration of movement, an increase in the depth of the summer active layer, and a decrease in the ice content, that, at some point, causes an acceleration of movement (Volume 2, Chapter 4).

Runoff: Annual runoff in Austrian streams and rivers will tend to decrease due to the temperature-related increase in evaporation. Regionally, a strong decrease in annual runoff is expected in the south of Austria. Austria-wide projections of runoff decreases are between 3 and 6 % by the middle of the 21st century and between 8 and 12 % by the end of the century. They vary according to the climate scenario selected and the respective projection model. How far these decreases will be compensated for or intensified by changes in precipitation is still unclear because of the high level of uncertainty in precipitation projections (Volume 2, Chapter 2).

A climate-induced shift in the seasonal runoff characteristics of Austrian streams and rivers is very likely. Low water levels during winter in the Alpine region will tend to rise due to an increase in winter temperature and an earlier start of snow melt. For summer run-off, a slightly decreasing trend is expected, which will be considerably more pronounced in the south (Volume 2, Chapter 2).

The maximum annual flood flow rates have increased in around 20 % of the catchment areas during the past 30 years. Small catchment areas north of the main Alpine ridge are particularly affected. Across Austria, winter floods have increased significantly more than summer floods. The influence of climate change on these events cannot be proven at present because the increased number of floods over the past decades still falls within the bounds of natural variability. In future a temporal shift in the occurrence of floods toward earlier spring floods and more winter floods, particularly in northern Austria, is expected (Volume 2, Chapter 2). The damage potential

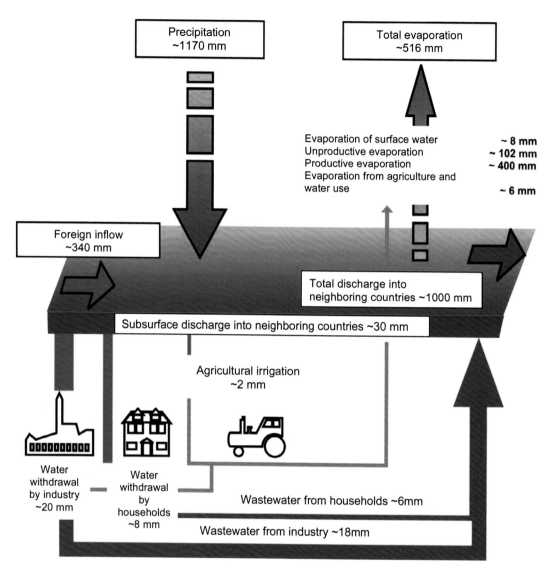

Figure S.2.4. Average values of the water balance for Austria during the period 1960 to 2000. Source: Central Hydrographical Buro, Austrian Federal Ministry of Agriculture, Forestry, Environement and Watermanagement, Dep. IV/4 Water Balance

of heavy precipitation in settlement areas – due particularly to the inadequate size of sewage networks, which do not have the capacity to absorb and discharge the volume of precipitation – is estimated to be high (Volume 2, Chapter 6). Due to the uncertain development of climate extremes (in particular, heavy precipitation), reliable projections of future changes in floods are not possible to date (Volume 2, Chapter 2).

In the past decades, **water temperatures** in lakes as well as streams and rivers have risen and are expected to continue to rise further. In the period from 2001 to 2005 lake temperatures during the swimming season (June to September) were 0.9 °C (Traun catchment area), 1.3 °C (Carinthian lakes), and 1.7 °C (Ager catchment area) higher than during the 1960–1989. The mean temperature increase across all measurement

stations in flowing waters since the 1980s is 1.5 °C in summer and 0.7 °C in winter (Volume 2, Chapter 2). In future, a further increase in water temperatures is expected, with lakes more affected than flowing waters. By the middle of the century, temperatures during the swimming season are expected to rise on average by between 1.2 and 2.1 °C in Carinthian lakes and between 2.2 °C and 2.6 °C in most Salzkammergut lakes (Figure S.2.3). In flowing waters an increase in temperature of 0.7 °C–1.1 °C in summer and 0.4 °C–0.5 °C in winter is expected by 2050 (Volume 2, Chapter 2).

Groundwater, soil moisture: In most areas of Austria, a decrease in ground water since the 1960s and a considerable increase in ground water since the middle of the 1990s have been observed. These fluctuations can largely be attributed to natural climate variability and changes in regional water use. Between 1976 and 2008, a decreasing trend in the annual average ground water levels was observed at 24 % of measuring stations, and an increasing trend at 10 % (Volume 2, Chapter 2).

In future, average soil moisture content and groundwater renewal will decrease moderately. While only very small changes in soil moisture are expected during the vegetation period until 2050, a slight decrease is expected for the months March to August during the period 2051 to 2080. Moreover, no large-scale change is expected in groundwater renewal until the middle of the century. Different climate scenarios yield different results with regard to groundwater renewal for the second half of the century; especially in the non-alpine areas, projections vary between +5 % and –30 %. For the south and southeast of Austria decreases are expected (Volume 2, Chapter 2).

Water balance: The water balance in Austria is currently characterized by higher water supply than demand. For the reference period from 1961 to 1990, average annual precipitation was between 1 140 and 1 170 mm (mm = liters per square meter), whereas industrial use was 20 mm, household use 8 m and agricultural water demand 2 mm (Figure S.2.4). In future, evaporation (currently 500–520 mm) is expected to increase, and runoff (currently 650–690 mm) is expected to decrease slightly. From a water management perspective, there is no real need to act until the middle of the 21st century, although areas with currently low water resources (particularly in the east and south of Austria) will need to adapt (Volume 2, Chapter 2).

Domestic water demand has decreased slightly in the past decades, and this trend is set to continue in future. This decreasing trend is due to more efficient water use in households and businesses and lower losses in water pipe networks. While the average Austrian household water consumption was 135 liters per person per day in 2011, specific consumption will decrease to approximately 120 liters per person per day by 2050 (Volume 2, Chapter 2).

The majority of agricultural water demand in Austria is rain-fed. However, in the east and in some locations in the southeast of Austria, water for irrigation is already needed now: groundwater and to a lesser extent surface water is being used. As a result of rising temperatures the water demand of agricultural crops will also rise, which means that particularly in the **east and southeast, irrigation demand will increase in the long run** (Volume 2, Chapter 2). Salinization of soils could occur as a result of increased irrigation (Volume 2, Chapter 5).

S.2.3 Impacts on Topography and Soil

Topography is determined by long-term geomorphological forces, although short-term forces such as climate factors can be superimposed. The large Alpine valleys e. g., were essentially shaped by the ice ages during the past 400 000 years; they are currently undergoing many short-term topographical changes (e. g., through landslides) which are decisively influenced by current (and future) climate factors, particularly temperature, radiation, and precipitation. (Volume 2, Chapter 4).

In addition to natural geomorphic processes, topography in Austria is also strongly influenced by human activity. Society changes the natural frequency and magnitude of geomorphological processes such as mudslides and landslides through, for instance, land use change or surface modifications (e. g., drainage). Moreover, society also shapes and modifies the material environment, thereby changing process flows in the terrain (e. g., through the construction of infrastructure or expansion of land use). Society also steers geomorphic processes (e. g., through river engineering, slope drainage, and changing vegetation) and can also act as a catalyst (e. g., floods or snow avalanches caused by malfunctioning of protective structures). Climate-induced changes in geomorphological processes and topography take effect in parallel with human influences. The influence of human and climate factors take effect sometimes in a reinforcing manner, sometimes in a diminishing manner, and their effects are often asynchronous (Volume 2, Chapter 4; Volume 2, Chapter 1).

Increasing heavy precipitation, extended precipitation events, or increased warm air advection during snow cover can enhance susceptibility to landslides. In this context, the type of land cover (e. g., forest, arable land, grassland) is of particular significance. On the whole, human interventions (e. g., land use change) are considered to be of greater importance for future landslide events than climate change. In general, there are still high levels of uncertainty regarding the future developments of landslides (Volume 2, Chapter 4; Volume 2, Chapter 5).

It is assumed that the frequency and magnitude of earth slides and flows as well as debris flows will increase in future. In particular, local increases in thunderstorms or extended precipitation events could lead to a future increase in landslide and flow and debris flow activity. Furthermore, the climate-induced decrease in permafrost and the decline of glaciers (Volume 2, Chapter 2), could lead to an enhanced danger of mudslides and rock falls as unsecured material becomes uncovered (Volume 2, Chapter 4).

In high altitude areas influenced by permafrost, climate change will lead to an increase in rockfall, rockslide, and debris flow activity. For most areas – those which are permafrost-free – hardly any change in activity is expected. Generally, in past warm periods in Austria, a shift in maximum rockfall activity from spring to summer could be observed. To date, observations have shown no climate influence on deep-seated large-scale landslides, such as Bergsturz or large rock falls and slides (Volume 2, Chapter 4).

In areas above 2 500 m above sea level in Austria, which accounts for approximately 2 % of territory (1 600 km²), permafrost has to be reckoned with. It is estimated that **an increase in temperature of 1 °C can cause a retreat of the permafrost line by approximately 200 m** (Volume 2, Chapter 4). Thus, expected warming would reduce the body of permafrost in the Austrian Alps and large areas would be free of permafrost in the future (Volume 2, Chapter 2).

Solifluction (soil flow in periglacial regions) is a slow, downhill flowing movement of thawed topsoil on still frozen subsoil (Volume 2, Chapter 4). The retreat of permafrost due to increased warming in the Alpine region will reduce solifluction at lower altitudes and increase it in high altitude regions.

Due to the retreat of glaciers (Volume 2, Chapter 2) **erosion and consequent sediment-input into flowing waters will increase in the areas that become exposed**. A direct consequence of glacial erosion is high sediment concentration in flowing waters and sedimentation in lakes, which in the latter case can lead to land aggradation. On the one hand, the local withdrawal of glaciers and the thawing of permafrost may lead to a significant increase in the potential sediment load and thus of the solid load discharge into flowing waters. On the other hand, it must be assumed that the complete disappearance of local glaciers will lead to a decrease in sediment load in water in the medium term. Furthermore, it is important to note that the transportation of sediment load in flowing waters is also significantly influenced by human interactions such as river regulation and water reservoirs used for drinking water supply, irrigation purposes, or power plants (Volume 2, Chapter 4).

If wind speeds increase locally, wind erosion could increase in future. However, this is also heavily dependent on (changes in) vegetation and agricultural use (Volume 2, Chapter 4; Volume 2, Chapter 5).

Climate-induced changes in geomorphic processes and consequently topography have only a minor effect on ecosystem services. As geological processes react more slowly to a changing climate than ecological processes (Volume 2, Chapter 3), the former will play only a minor role in the provision of ecosystem services (Volume 2, Chapter 1) in Austria during the decades to come (Chapter 2, Volume 4).

As far as soil is concerned, the most evident climate effects are expected to be on soil life and consequently on humus accumulation. Soil is described as the uppermost part of the earth's crust affected by weathering, in other words, the upper decimeters that are in direct interaction with the atmosphere. Many soil processes are dependent on both temperature and moisture – their future development therefore depends on local changes in both temperature *and* precipitation. Particularly affected are soil life and processes of humus decomposition, nutrient availability, and possible changes in soil structure (Table S.2.1). Dry soils tend to have less diversity of soil life and less robust populations than moist soils with a good oxygen supply (Volume 2, Chapter 5).

On the whole, soil reacts slowly to changes in climate. However, as vegetation reacts considerably faster to changes in climate (Volume 2, Chapter 3) and co-determines soil development – particularly the development of organic substances in soil – indirect climate effects (Volume 2, Chapter 1) can be expected to have the main impact on soils in the short- and medium term (Volume 2, Chapter 5).

Soils influence the carbon budget and consequently have a direct effect on the climate. The amount of CO_2 entering the atmosphere from soils each year (roughly equal amounts are taken up by soils again), considerably exceeds the emissions caused by fossil fuels. Preserving the ecological budget of soils is thus an important aspect of climate protection. Higher temperatures increase mineralization and can lead to a decrease in organic substances and consequently in the stored carbon in soils. However, a prerequisite for this is constant moisture conditions. Dry periods delay humus mineralization as does freezing of the ground during thick snow packs (Table S.2.1). Site and land use determine if, and to what extent, the humus expected to be lost due to rising temperatures can be compensated for by increased biomass production of vegetation (e. g., by increased levels of CO_2 and longer vegetation periods, Volume 2, Chapter 3); this still shows high levels of uncertainty. There is also a research gap regarding the stability of humus complexes and the role of subsoil in carbon storage (Volume 2, Chapter 5).

Temperature extremes and dry phases have greater effects on soil processes than gradual climatic changes. Temperature extremes influence, for example, soil biota more than gradual changes in average temperature do. Temperature extremes and dry phases also have a strong influence on turnover rates of carbon and nitrogen in soils. They increase during strong and lengthy freezing and thawing processes in winter

Table S.2.1 Assessment of the sensitivity of processes in soils related to climate change. Developed by Geitner for AAR14

Processes	Sensitivity	Explanations
Related to the mineral constituents		
Physical weathering	+ +	D or I: depending on elevation zone (frequency of freeze / thaw cycles)
Chemical weathering	+ +	I: with temperature increase (nival / alpine zone) D: under dry conditions
Biological weathering	+	D or I: with vegetation changes
Oxidation	+	I: under dry conditions
Reduction	+	I: under wet conditions
Clay mineral formation	+	D: under dry conditions
Clay displacement	+	D: under dry conditions
Podsolization	+	D: under dry conditions
Calcification	+	I: under dry and alternate wetting and drying conditions D: under wet conditions
Related to the organic constituents		
Mineralization	+ + +	I: under average conditions D: under dry or very wet conditions
Humification	+	D or I: Depending on further factors (e. g. wetness, chemical composition of litter)
Others		
Exchange processes (ions)	+	D: under dry conditions
Aggregate formation	+	Depending on other conditions
Bioturbation	+ +	Depending on other conditions
Cryoturbation	+ +	Depending on duration of frost periods and frequency of freeze / thaw cycles, according to elevation

D = decrease, I = increase, + = moderate, + + = average, + + + = strong effects expected

(through changes in duration and depth of snow cover) and also in the case of intense and lengthy drying out of soils, followed by heavy precipitation events; this leads to spikes in GHG emissions directly after such events (Volume 2, Chapter 5).

Drier conditions could bring about local reduction in seepage water and groundwater level (Volume 2, Chapter 2) on water-influenced soils (gley, pseudogley), reducing waterlogging and increasing yields. However, such changes could also impair the natural dynamics of floodplain and bog soils. Bog soils, which are a significant carbon reservoir, react in a particularly sensitive way to increasing temperatures and desiccation (Volume 2, Chapter 5).

An increase in climatic extremes affects arable soils more than grassland soils. Erosion of arable land through water and wind can particularly increase during phases of incomplete or poor vegetation cover (Volume 2, Chapter 4; Volume 2, Chapter 5). In future, cultivation techniques and soil management will become increasingly important and will need to be adapted to compensate for potential climate-induced problems. Grassland soils are expected to be more stable, although they may also be subject to a climate-induced reduction in humus (Volume 2, Chapter 5). In light of the global questions about food security and concurrently rising demands on soils (e. g., through increased use of bioenergy crops), an increase in nitrogen utilization efficiency in soils could make a positive contribution (Volume 2, Chapter 5).

Naturally balanced soils will fulfil their functions and services under changing climatic conditions better than soils that have been subject to intense human degradation. Soil protection is therefore not just a means of climate protection, but also an important contributing factor to climate adaptation.

Rising temperatures lead to increased CO_2 emissions from forest soils. A temperature increase of 1 °C leads to approximately 10 % more CO_2 being emitted through soil respiration; a temperature increase of 2 °C leads to approximately 20 % more CO_2 and N_2O emissions. An increase in disruptions (e. g., through windbreak events and following bark-beetle infestations (Volume 2, Chapter 3) also leads to humus and soil loss through erosion, which again leads to increased CO_2 emissions from soil and also to an impairment of hydrological soil functions (Volume 2, Chapter 5).

To date, fairly little is known about impacts of climate change on high alpine and urban soils. Climate-induced change in vegetation – particularly in the region of the current tree line (Volume 2, Chapter 3) – would also influence humus quantity and quality in high alpine soils. On the other hand, higher temperatures would enhance humus depletion, particularly as the organic matter of soils at high altitudes contains easily degradable components. However, due to the small-scale differentiation of mountain soils, there is a limited number of general statements that can be made (Volume 2, Chapter 5). It is assumed that urban soils are highly at risk of climate change, as the urban environment per se subjects them to increased temperatures and reduced water content and natural soil structure is often missing. However, there are no detailed studies on the climate sensitivity of urban soils in Austria (Volume 2, Chapter 5).

S.2.4   Impacts on the Living Environment

In areas with low levels of precipitation, such as north of the Danube and in eastern and south-eastern Austria, agricultural yields will decrease. In the cooler areas of Austria with higher levels of precipitation, however, a warmer climate will increase the potential yields of agricultural crops. Rainfed summer crops with low temperature demands, such as spring cereals, sugar beet, and potatoes will be increasingly subject to heat stress and drought damage. The potential yield of these crops will stagnate or decline, particularly on light soils with low water storage capacity (Volume 2, Chapter 3). It is possible that currently rainfed crops will increasingly need to be irrigated in certain regions (Volume 2, Chapter 2; Volume 2, Chapter 3). In irrigated areas the water demand for irrigation will increase (Volume 2, Chapter 3).

The yield potential of thermophilic summer crops such as maize, soya beans, and sunflowers could increase, as long as there is sufficient water supply. It should be noted, that the increased yields during the past decades are primarily due to progress in agricultural technology, agro-chemical measures, and plant breeding – and not to climate change. However, in Austria and Switzerland the inter-annual variability of yields has increased, which can at least in part be attributed to climate change (Volume 2, Chapter 3).

Winter crops could also experience an increase in yield potential, as they make better use of winter moisture in soils. However, in wet locations or regions with high levels of precip-

itation there is danger of waterlogged soils because of increasing precipitation in winter. Winter crops (e.g., winter wheat) are also at an increased risk from pests and diseases due to warmer winters (Volume 2, Chapter 3).

A further increase in temperature will favor wine cultivation in regions of Austria that currently have less suitable climate conditions. In the current wine cultivation areas, an increase in temperature will be particularly favorable for red wine varieties and the quality of red wine. For white wine, where acid content is an important quality feature, quality could improve in colder or new cultivation areas, but could also decrease in current cultivation regions (Volume 2, Chapter 3).

Fruit crops will be negatively affected by the expected climatic changes. Increased aridity and irrigation demands will be particularly problematic, as fruit crops need much water and are more sensitive to heat and drought than vines, for example. An increase in the amount and intensity of thunderstorm activity could increase the danger of crop damage, particularly hail. In valleys and basins increasing late frost damage, particularly during flowering, is expected. Furthermore, phases of extreme weather conditions could cause disruptions in growth rhythms. Warmer weather conditions in winter, for example, could lead to a decrease in the frost resistance of fruit trees, increasing the danger of damage during the next frost (Volume 2, Chapter 3).

Farm animals also suffer from climate change. Increasing heat periods can decrease the performance and increase the risk of disease in farm animals. Increasing heat stress can lead, for example, to a decline in the milk production of cows or to a decrease in the size of hens' eggs. Next to ambient temperature, humidity and airflow also influences the thermal wellbeing of animals (Volume 2, Chapter 3).

The productivity potential of Austrian forests in mountainous regions and in regions with sufficient precipitation will increase due to the expected changes in climate. In contrast, productivity in eastern and north-eastern areas at low altitudes and in inner-Alpine basins will decrease due to an increase in dry periods. Whether potential increases in growth can actually be achieved in managed forests will depend largely on the numerous risk factors and the changes in them induced by climate (Volume 2, Chapter 3). For example, it is expected that the intensity and frequency of climate-related disturbances in forest ecosystems will increase under all warming scenarios. This is particularly true for occurrences of thermophilic insects such as bark beetles. Furthermore, new types of damages through imported or harmful organisms migrating from southern regions can be expected. Abiotic disturbance factors such as storms, late and early frosts, and wet snow events may also occur more frequently than previously, which in turn will amplify damage from biotic disturbances (e.g., by bark beetles). Intensified disturbance regimes lead to lower wood production revenues and can impair other ecosystem services such as the protective function against, for example, rock fall, landslides and avalanches, and also the carbon storage capability of forests (Volume 2, Chapter 3; Volume 2, Chapter 4).

The dry summers of 2003 and 2007 showed that under certain weather conditions **forest fires** can also develop rapidly and reach significant dimensions in Austria. Due to the expected warming and increasing likelihood of dry weather periods in summer, an increase in the frequency and severity of forest fires is expected in the Alpine region in the future. Fire is a particular risk for the Alpine region because of the long regeneration time needed by vegetation following forest fires and the fact that the latter can reduce or eliminate the protective function of forests against natural hazards (Volume 2, Chapter 4; Volume 2, Chapter 3).

In Austria the competitive strength of deciduous forests will increase. Up to a temperature increase of 1 °C, the distribution of forest communities would not change significantly; however, a warming of 2 °C would lead to a change in the potential natural vegetation. This means that in close to 80 % of Austrian forest area, without forest management, the potential habitat of beech-, oak-, and beech-fir-spruce forest types would increase (Volume 2, Chapter 3).

In alpine regions, plants that are adapted to cold conditions will be displaced by thermophilic species and they themselves will advance to higher altitudes. The upward shift of species in all alpine summit regions has been verified across Europe and could lead to a temporary increase in biodiversity in higher alpine regions in Austria. Despite warming, cold-adapted vegetation still manages to survive in niches, however in the medium term local extinction of cold-adapted species in alpine vegetation can be expected (Volume 2, Chapter 3).

Peat bogs are heavily affected by climate change. It is estimated that 85 % of peat bogs in Austria will be endangered by a temperature increase of around 2–3 °C (Volume 2, Chapter 3).

Climate change will affect Austrian fauna. This can already be seen in species shifts of various animal groups, such as dragonflies, beetles, and invertebrate freshwater animals. At the same time, climate-induced habitat loss is expected in many animal species and species groups, including many endemic species (i.e. species that do not occur elsewhere). Species habitat shifts not only depend on the reaction of the respective species to climate change, but also on the ability of the species to migrate and establish itself against other species that already live in the new habitat (Volume 2, Chapter 3).

Amphibians, due to their specific habitat demands and low mobility, are particularly vulnerable to climate change. Indirect impacts of climate change (Volume 2, Chapter 1) such as habitat loss, for example, a possible periodical decrease in small water bodies, and the loss of wetlands following more frequent or prolonged dry periods (Volume 2, Chapter 3) are of highest relevance (Volume 2, Chapter 3) In this context, projected changes in precipitation distribution are probably a greater risk factor than changes in temperature (Volume 2, Chapter 3).

Reptiles are potential winners of climate change. Longer summer conditions will mean an increase in reproductive success for reptiles. Successful reproduction of non-native reptile species (such as species of turtle) in the wild has already been observed on occasion (Volume 2, Chapter 3).

A shift toward warm-water-loving fish species is expected. Warming of 2.5 °C (Volume 2, Chapter 2) could result in an altitudinal shift of fish regions by 70 m and relocation of fish regions upstream by approximately 30 km (Volume 2, Chapter 3). However, this theoretical relocation upstream will not be possible in many cases, as upstream habitats are often not suitable for the fish. Accordingly, a loss of trout and grayling waters can be expected overall. Over half of native fish species are already on endangered species lists; the additional pressures of climate change and also continued expansion of hydropower will further endanger native fish fauna (Volume 2, Chapter 3).

Climate change not only impacts individual plant and animal species, but also strongly influences their interaction in ecosystems. The relationship between predator and prey, parasite and host, and plant and pollinator could change as a result of changes in future climate. A temporal decoupling of processes, such as the flowering time of plants and stage of development of pollinators or a geographical "drifting apart" of habitats (low overlap of habitats of interacting species in the future) could have a strong influence on ecosystems (Volume 2, Chapter 3).

On the whole, ecosystems with long development periods are particularly affected by climate change. In Austria these include forests, habitats above the tree line, and moorland. Because of these slow development periods relative to climate change, the ability of such ecosystems to adapt to changes in climate is limited. The resulting climate vulnerability affects both plants and animals in such ecosystems (Volume 2, Chapter 3).

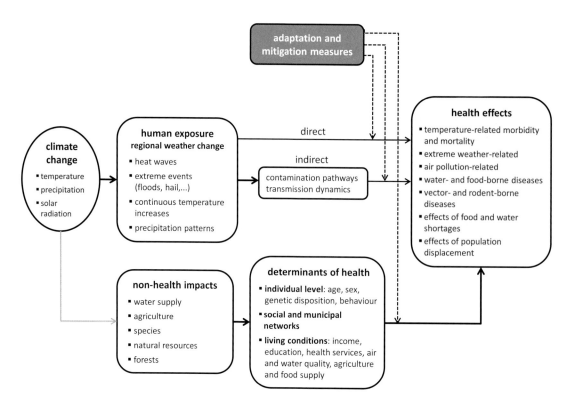

Figure S.2.5. Direct and indirect impact chainsinfluencing pathways of climate change affecting health. Source: adapted from Confalonieri et al. (2007); McMichael et al. (2004)

S.2.5   Impacts on Humans

The increase in heat waves leads to rising mortality rates.
Temperature increase – in particular heat waves – probably has the most serious direct effects on human health (Figure S.2.5). Heat burdens the human organism and can lead to death, particularly where health is poor. Cardiovascular problems, especially in older people, but also in infants or the chronically ill, are increasingly being observed, particularly following dehydration. A regionally dependent temperature exists at which the mortality rate is at its lowest; beyond this temperature mortality increases by 1–6% for every 1 °C increase in temperature (very likely, high confidence, Volume 2, Chapter 6; Volume 3, Chapter 4). In particular, older people and young children have shown a significant increase in the risk of death above this optimum temperature. To date, little is known about adaptation possibilities and speed of adaptation to higher mean temperatures. Heat waves particularly affect people in urban areas as they are intensified by the urban heat island effect (higher turnover of radiation energy and heat accumulation) and potentially prolonged as well as enhanced in large cities. Nocturnal cooling is also considerably lower in urban than in rural areas, which affects nocturnal recovery phases (Volume 2, Chapter 6). During the heat wave in 2003, between 180 and 330 heat-related deaths were registered in Austria. Precipitation-induced extreme events (floods, landslides etc.) increase the risk of injury or death. The risk of epidemics, often associated with floods in emerging or developing countries, is less of a problem in Austria due to high levels of hygiene.

Indirect climate impacts on human health due to the spread of non-endemic animal and plant species is expected. Pathogens transferred by blood-sucking insects and ticks play a particularly important role, as not only the agents themselves, but also the vectors' (insects and ticks) activity and distribution are dependent on climatic conditions. Newly introduced pathogens (viruses, bacteria and parasites, and also allergenic plants and fungi such as, for example, ragweed (*Ambrosia artemisiifolia*) and the oak processionary moth (*Thaumetopoea processionea*), and new vectors like the tiger mosquito (*Stegomyia albopicta*) can establish themselves, and existing pathogens can spread (or disappear) regionally. A shift in the tick population to higher regions can already be observed. Rodents act as important vector carriers and reservoirs, and their distribu-

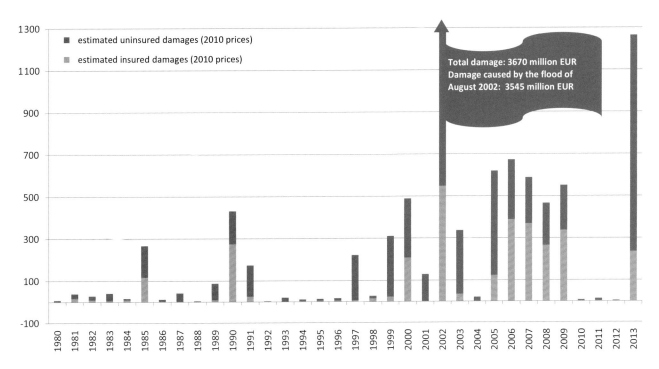

Figure S.2.6. Weather and climate related damage in Austria 1980 to 2010. Copyright: Munich Re Geo Risks Research, NatCatSERVICE (2014)

tion and populations shifts with climate change. As complex predator-prey relationships are involved in this process, it is difficult to make concrete predictions about future developments. Among the illnesses that might increase in frequency are meningitis, transmitted by ticks and other vectors, yellow fever, dengue, malaria, leischmania, hanta virus, and flu-like illnesses. Pathogens that are also transmitted via drinking water and food (e.g., salmonella) are temperature-dependent and can spread further as a result of higher average temperatures (Volume 2, Chapter 6).

Health effects due to climate change are closely linked to social conditions. Usually the coincidence of several factors (e.g., low income, low level of education, low social capital, precarious working and living conditions, unemployment, limited possibilities to take action) makes less privileged population groups particularly vulnerable to climate change impacts. Poorer people are particularly vulnerable to climate change due to the location of their apartments and houses in settlement areas (e.g., in dense areas with low levels of green areas, areas at risk of flooding) and in particular due to the structural condition of the buildings they live in; poorer people also have fewer possibilities of adaptation (e.g., to increasing heat waves) and health protection. In the face of rising energy prices, the weakening and shortening of cold seasons (fewer heating days) can be seen as a relieving factor for vulnerable social groups (Volume 2, Chapter 6).

Although women frequently act in a more climate friendly manner than men, they are often – even in Austria – more affected by climate change. During the heat wave in 2003, considerably more women died than men (across all age groups) in Europe (Volume 2, Chapter 6).

Climate-induced migration pressure on Austria from developing and emerging countries will increase. This is largely due to the global imbalance between polluters, namely, the leading per capita GHG emitters (industrialized countries) and the people most vulnerable to and affected by climate impacts (developing countries) (Volume 2, Chapter 6; Volume 2, Chapter 1). Whether increased migration pressure will lead to a higher number of immigrants will, however, depend on political responses (Volume 2, Chapter 6).

Compared to the rest of Europe, direct climate impact costs close to the average are expected for Austria (Volume 2, Chapter 6). The expected high variability in precipitation will primarily affect agriculture and forestry and, to a lesser extent also, energy and water management in eastern and south-eastern parts of the country (Volume 2, Chapter 2; Volume 2, Chapter 3). The potential increase in extreme precipitation events and their indirect consequences, such as floods and landslides or avalanches, have high damage potential for infrastructure, particularly in the alpine region, hilly regions and in several river valleys (Volume 2, Chapter 2; Volume 2, Chapter 4). It is important to note that an absolute estimate of climate impact costs is difficult, as there needs to be a consideration not only of changes in the quality and quantity of goods and services but also of the effects on ecosystem services which have no "market value" as such (Volume 2, Chapter 6).

Total economic damage due to extreme weather events has strongly increased in Austria in the past years. On the basis of insurance data, total damages due to extreme weather events in Austria between 1980 and 2010 are estimated at € 10.6 billion (in 2013 prices, Figure S.2.6). It should be noted that both the quantity and intensity of weather events and the increasing exposure of assets to extreme weather events are responsible for the growing economic costs. Moreover, due to strong settlement growth in risk regions, the damage potential has risen considerably in the past decades.

During the period 2001 to 2010, certain events were particularly cost-intensive: the floods in 2002 (around € 3.7 billion), 2005 (almost € 0.6 billion), and 2013 (almost € 0.7 billion), with a number of heavy winter storms each incurring damage of several hundred millioneuros. Note that these figures include only the costs of direct damage incurred through reconstruction and repairs; the indirect costs of the knock-on effects of such weather events were not considered. Furthermore, much damage from small and slow-onset events (e. g., droughts) is not considered. This means that total economic damages from weather events should be considerably higher than the values cited here (Volume 2, Chapter 6).

It is highly likely that winter tourism in Austria will be negatively affected by climate change. Winter warming and the associated shortening of the season and disruptions to it, as well as the low snow reliability at lower altitudes and in eastern parts of the country, will negatively affect winter tourism in its current form. Furthermore, increased dependence on water- and energy-intensive artificial snowmaking can be expected in all regions of Austria (Volume 2, Chapter 6).

Both spa and recreational tourism could profit from increasing temperatures and lower precipitation frequency in future. Summer tourism especially will profit from climate change and Austria could position itself as a "summer freshness" resort particularly for Mediterranean countries. However, this potential would have to be utilized accordingly, and it is not yet clear how far these opportunities will materialize in future. Increases in summer tourism are unlikely to be able to compensate for losses in winter tourism (Volume 2, Chapter 6).

Altogether, city tourism appears to be fairly robust in climate change terms. Effects are expected inasmuch as the activities of city tourists will center more on green areas, parks, and gardens and courtyard restaurants, while non-air-conditioned buildings could be avoided in summer. Furthermore, a shift in the number of visitors from summer to spring and autumn is expected (Volume 2, Chapter 6).

With regard to energy demand, climate change-induced energy savings on heating will most probably significantly surpass additional energy demand for cooling (Volume 2, Chapter 6). In Austria, around 60 % of electricity demand is covered by hydropower. In future, a slight reduction in hydropower production is expected, and production will decrease in summer and increase in winter, due to climate. Current projections vary in their estimates of changes in annual production by the end of the century from ±5 % to –15 %. With regard to cooling water demand for power plants, regional and seasonal constraints are possible, for example, in summer for catchments without any glacial / nival buffers. Thermal plants located on larger rivers (Drau, Inn, Mur, Danube) should not be subject to any future usage constraints (Volume 2, Chapter 2).

Settlement areas that are not endangered by natural hazards will shrink. Currently around 400 000 buildings in Austria are located in flood-endangered regions. Settlement areas are expected to continue to expand into flood-endangered regions, unless restrictions are imposed by planners (Volume 2, Chapter 6). Furthermore, a climate-induced expansion of flood zones must be expected (Volume 2, Chapter 2). Expansions of settlement areas will be complicated, particularly in Alpine valleys, as there will be an expansion both in flood zones in the valleys and on slope areas that are endangered by landslides (Volume 2, Chapter 2; Volume 2, Chapter 4; Volume 2, Chapter 6).

Energy and transport infrastructures demonstrate a high level of exposure to climate change, particularly as they are often located in exposed areas. Due to the network structure, an interruption at a single point can often lead to large-scale service disruptions. Transport infrastructures are very likely to be affected by extreme precipitation events – currently more than three-quarters of all damage – from geomorphic processes triggered by extreme precipitation (e. g., landslides including earth flows and slides, rock falls and debris flows, undercut of river banks, snow avalanches affect transport infrastructure. The extent of direct damage as a result of future extreme events depends on the scenario; in any case, the indirect damages and related costs can be expected to be considerably higher than the direct costs (Volume 2, Chapter 6).

Through cascading effects, weather-induced disruptions to energy infrastructure can lead to large-scale "black-outs." On the one hand, danger comes from the physical damage caused by landslides and floods (Volume 2, Chapter 2; Volume 2, Chapter 4; Volume 2, Chapter 6). Heat waves can lead to network problems, and heat can cause problems in energy production (low water, reduced supply of cooling water, reaching allowed temperature limits) and to trans-alpine transmission lines toward Italy (high energy demand and low power plant productivity in southern Europe), which are particularly strained by simultaneously rising energy demand in Austria (cooling energy and irrigation) (Volume 2, Chapter 6).

S.3 Climate Change in Austria: Mitigation and Adaptation

S.3.1 Climate Change Mitigation and Adaptation

Global emission reduction requirements. Global GHG emissions to date continue to rise along the path of the "business-as-usual" (BAU) scenario. If this trend continues, emissions will have doubled by mid-century. Stabilizing global annual mean temperature increase below 2 °C by the end of the century (compared to pre-industrial levels) will require global GHG emissions reductions of at least 50 % by mid-century compared with present levels – and up to 90 % in industrialized countries (Volume 3, Chapter 1).

The change in the annual global mean temperature of 4 °C and above, which is to be expected in a BAU scenario, is equivalent to the transition from the ice age to the interglacial period. Compared to the past 10 000 years in which human civilizations have emerged, a 4 °C warmer planet will have consequences for nature and humanity that will be almost impossible to control. Although warming of 2 °C would also bring significant changes, it can be seen as a threshold for the avoidance of catastrophic consequences (Volume 1, Chapter 1; Volume 3, Chapter 1).

Both mitigation and adaptation measures are essential for any given global temperature stabilization level. Mitigation of

GHG emissions requires both technological change, such as, for example, efficiency gains, and behavioral change to reduce resource use and the resulting emissions per unit of activity. The aim is to reduce climate change by managing its driving forces. Adaptation to climate change on the other hand, includes initiatives and measures that reduce the vulnerability to, or increase the resilience of, human-environmental systems against acute or expected impacts of climate change, for example, flood prevention measures or the cultivation of better adapted plant species.

Determined and vigorous emission mitigation measures are necessary to reach any climate stabilization target. Complete implementation of the voluntary emission reduction pledges specified in the Cancun and the Copenhagen Accords correspond to a path that will lead to a global warming of more than 3 °C (with a 20 % likelihood of more than 4 °C) by the end of the century (see Figure S.3.1).

On a global level there is significant mitigation potential in energy production, transportation, buildings, industry, agriculture, forestry, and waste management. These are described in the respective chapters of the full AAR14 Report (see Volume 3) in more detail.

Within the framework of the European "Burden Sharing Agreement" in implementing the Kyoto Protocol, Austria committed to reducing GHG emissions by 13 % (period 2008 to 2012, relative to 1990 levels). However, and in contrast to the majority of other EU member states (including Germany, the United Kingdom, France and Sweden), GHG emissions in Austria increased considerably. Consequently, Austria was unable to fulfil its Kyoto targets by domestic emission reductions. Formal compliance was achieved with the purchase of about 80 Mt CO_2-eq. of emission permits on the global market at a cost of roughly € 500 million.

The central pillar of European climate policy is the European Climate and Energy Package, which compris-

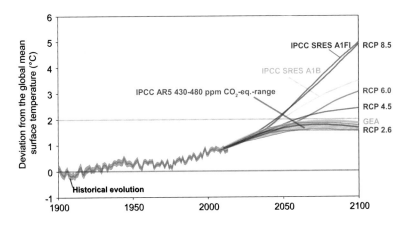

Figure S.3.1. Global mean surface temperature anomalies (°C) relative to the average temperature of the average of the first decade of the 20th century, historical development, and four groups of trends for the future. Two IPCC SRES scenarios without emission reductions (A1B and A1F1), which show temperature increases to about 5 °C or just over 3 °C to the year 2100, and four new emission scenarios, which were developed for the IPCC AR5 (RCP8, 5, 6.0, 4.5 and 2.6), 42 GEA emission reduction scenarios and the range of IPCC AR5 scenarios which show the temperature to stabilize in 2100 at a maximum of +2 °C Sources: IPCC SRES (Nakicenovic et al. 2000; IPCC WG I 2014 and GEA 2012)

es the three central goals of i) reducing of GHG emissions by 20 %, ii) increasing the share of renewables in final energy consumption to 20 %, and iii) increasing energy intensity by 20 % ("20-20-20 targets") by 2020 in relation to 2005. The Climate and Energy Package is supplemented by a range of further measures and directives (European emissions trading, energy efficiency, promotion of renewable energies, eco design, energy performance of buildings, combined heat and power; Volume 3, Chapter 1; Volume 3, Chapter 6).

In February 2011 the European Council approved the plan to reduce the European Union (EU) GHG emissions by 80–95 % by 2050 (compared to 1990 levels) in order to limit dangerous climate change and keep the average increase in temperature below 2 °C (compared to pre-industrial levels). Several European countries (the United Kingdom, Denmark, Finland, Portugal and Sweden) have committed themselves to concrete emission reduction targets for 2050. **To date, Austria has set only short-term climate and energy reduction targets for the period up to 2020** (Volume 3, Chapter 1; Volume 3, Chapter 6).

The measures taken so far in Austria are insufficient to meet the commensurate contribution expected from Austria to achieve the global 2 °C stabilization target. To achieve targets set in the EU climate and energy package by 2020, it is estimated that Austrian emissions need to be reduced by 14 Mt CO_2-eq. compared to a reference scenario. This comparably small reduction could be reached, for example, by implementing a package of measures for technological options focusing on energy efficiency, which would require an annual investment volume of € 6.3 billion in the period 2012 to 2020. In addition to the mitigation effect, these measures would result in an increase in economic output by approximately € 9.5 billion and would generate around 80 000 additional jobs. At the same time the resulting energy savings would amount to € 4.3 billion in 2020 (using conservative assumptions regarding future energy prices; Volume 3, Chapter 1).

Austria has a considerable shortfall with respect to energy intensity improvement. In contrast to the EU average, energy intensity has remained relatively constant in Austria during the past two decades (energy use per unit GDP; see Figure S.3.2). In comparison, energy intensity in the EU28 has declined by 29 % since 1990 (e. g., by 23 % in the Netherlands, 30 % in Germany, and 39 % in the United Kingdom). However, at least a part of the improvements in Germany and the United Kingdom can be attributed to the relocation of energy-intensive production to other countries. With regard

to emission intensity (GHG emissions per unit of energy) Austria, where improvements reflect the growth in renewable energy since 1990, and the Netherlands are the countries with the largest improvements. Together, these two factors determine the GHG emissions intensity of GDP, which has declined in both Austria and the EU28 as a whole since 1990. In other words, GHG emissions have grown at a slower rate than GDP. The comparison with the EU28 shows quite clearly that Austria has to catch up considerably in energy intensity terms (Volume 3, Chapter 5).

Climate change causes high costs at both the global and European level. Global damages attributed to climate change are well beyond € 100 billion per year and could even be beyond a trillion per year. In Europe, costs of damages due to extreme weather events in 2080 are estimated at between € 20 billion (for a global warming of 2.5 °C) and € 65 billion (for a global warming of 5.4 °C and large sea-level rise). However, these cost estimates are subject to many uncertainties and do not include components that are difficult to quantify in monetary terms such as, for example, the loss of unique habitats. As for most countries, detailed studies on the costs of climate change in Austria to date are available only for selected sectors and regions (Volume 3, Chapter 6).

Despite existing uncertainties as to the specific extent of climate change impacts for different regions and sectors, early **planning and implementation of specific adaptation measures** is crucial. Any delay reduces the options for successful adaptation and increases related costs. **Adaptation measures can alleviate the negative impacts of climate change somewhat, but they cannot fully offset them** (medium confidence). For foresightful adaptation planning and implementation a broad range of measures can be taken by affected citizens, municipalities / regions, and at the federal level, or by private and public institutions; these include capacity building, technology-oriented measures, or changes in cultivation (Volume 3, Chapter 1; Volume 3, Chapter 6).

Initially "National Adaptation Programs of Action" (NAPA) were the primary supporting instrument for the states most vulnerable to climate change at the international level, under the auspices of UNCCD (starting in 1994) and UNFCCC (since 2001). Without implementation of adaptation measures, climate change-induced damages in developing countries are roughly estimated to amount to between € 25 and 70 billion in 2030. In contrast, current cumulative financial support from industrialized countries for adaptation in developing countries under the UNFCCC amounts to less than € 0.8 billion (Volume 3, Chapter 1).

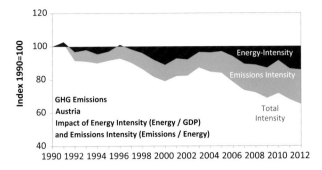

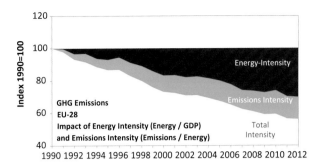

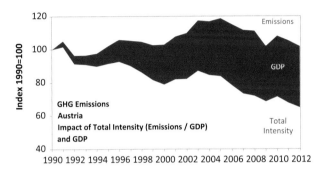

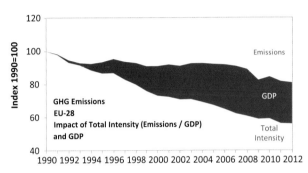

Figure S.3.2. Development of the GHG emission intensity of the GDP and the embedded relative importance of energy intensity (energy use per PJ GDP) in Austria and the 28 member states of the EU (upper panel). When combining this GHG emission indicator per GDP with the clear upward development of the GDP (lower panel), Austria shows an increase of GHG emissions during that period (+5 %) while emissions dropped in the EU-28 (−18 %). Source: Schleicher (2014), based on Eurostat

Adaptation has been on the agenda at the European level since 2005 and has been integrated into the Second European Climate Change Program (ECCP II). With its green and white books on adaptation, the European Commission (EC) has taken the first steps toward increasing the resilience of the EU to climate change. While the green book argues for the necessity of adaptation at the European level, the white book presents a framework for action within which the EU and its member states should prepare for the impacts of climate change. The EU adaptation strategy was enacted in spring 2013 (Volume 3, Chapter 1).

European activities at the political level, such as the publication of the green and white books on adaptation, but also new insights from research have prompted a number of European states to develop national climate change adaptation strategies. To date 14 European countries (Belgium, Denmark, France, Germany, Hungary, Malta, the Netherlands, Norway, Austria, Portugal, Switzerland, Spain, and the United Kingdom) have enacted an adaptation strategy. **In 2012 Austria adopted a national adaptation strategy specifically to cope with the consequences of climate change.** The effectiveness of this strategy will have to be evaluated in principle by how successful individual sectors, or rather policy areas, will actually be in developing appropriate adaptation strategies and in their implementation. An evaluation, for instance by regular surveys of the effectiveness of adaptation measures, already implemented by other countries, does not yet exist in Austria (Volume 3, Chapter 1).

Overall, studies on the costs of climate adaptation measures both in Europe and in Austria cover selected sectors and regions. Consequently, cost ranges are large, and further research is required to provide a better basis for cost / benefit estimates.

As both mitigation of and adaptation to climate change are required, coordination is needed, for instance with regard to the different time frames involved. Insufficient mitigation leads to a need for massive adaptation, which increasingly cannot be managed with "soft" or "green" measures, but will rather require grey / technical and more cost-intensive measures. Conversely, as adaptation measures can be CO_2-emissions intensive, these activities need to be coordinated with mitigation measures, so that they are supported rather than counteracted (Volume 3, Chapter 1).

The long useful life spans of infrastructural installations can lock in emission-intensive development paths for decades (lock-in effect). Investments in production processes, transport systems, energy use, and transformation should be screened with regard to lock-in effects, as existing capital stocks impede and increase the costs of mitigation measures over the entire useful life (Volume 1, Chapter 5; Volume 3, Chapter 1; Volume 3, Chapter 6).

Beyond creating an environment that is conducive to transformation, eliminating barriers is a core and crucial field of activity. This topic is also gaining importance internationally, and is embedded in theoretical-conceptual discussions on developing an appropriate framework for an environment that is conducive to transformation.

It is a fact that, despite well-founded studies on climate impacts, appropriate action to protect the climate and to adapt to climate change has not yet been taken, either internationally or in Austria. This is attributable in particular to barriers. In Austria, the following barriers have been identified (high confidence).

1. *Institutional barriers:* Due to the complex sectoral and federal split of competences, existing administrative structures are unsuitable for effectively dealing with climate change. The short time horizons of elected political decision makers – short relative to the comparatively slow, but steady processes of climate change – are also a barrier. International framework conditions also play an important role.

2. *Economic barriers:* In many individual economic decisions self-interest dominates collective wellbeing. When climate impacts are not, or only insufficiently, factored into prices or market rules, markets fail to solve the problem of climate change. Furthermore, so-called *rebound*-effects can take effect when cost savings resulting from increased energy efficiency lead to higher energy demand.

3. *Social barriers:* Households and companies demonstrate a discrepancy between environmental awareness and action actually taken. This is often due to a lack of confidence that individual action can make a relevant contribution on the aggregate level.

4. *Uncertainty and insufficient knowledge:* Differing opinions on reciprocal influences between natural, technical, and social systems (e.g., the extent to which technological options can solve the climate problem) as well as contradictory news coverage, dampen the willingness for meaningful action.

Examples of approaches for overcoming these barriers include a comprehensive administrative reform to make it fit for the purpose or the formation of new price structures in which the costs of products and services reflect their climate change impacts, as well as corresponding regulatory frameworks, a stronger integration of people from, for instance, civil society and academia in decision-making processes, targeted increase

of climate and environmental knowledge, and the closing of knowledge gaps.

S.3.2 Agriculture and Forestry, Hydrology, Ecosystems and Biodiversity

Climate change is a particular challenge for the management, use, and protection of terrestrial and aquatic ecosystems and for the sustainable management of the key resource, water. These challenges vary from affected system to system – from almost exclusively natural ecosystems and protected areas to intensively used agro-ecosystems.

The land system is characterized by close links between social, economic, geomorphological, climatic, and ecological factors. **Numerous climate-relevant systemic feedbacks exist between agriculture and forestry, water management and protection, and the preservation of ecosystems and biodiversity.** Because of these systemic effects, changes in one area, such as economy and society, can impact many other areas (Figure S.3.3; Volume 2, Chapter 3; Volume 3, Chapter 2).

In this context a measure to reduce GHG emissions – for example, increasing forest areas and stocking density in order to sequester carbon (C) – can lead to (positive or negative) effects on i) productive capacity (such as agricultural and forest production) and other ecosystem services (such as water retention capacity or protection against avalanches or landslides), ii) biodiversity, iii) the risk of damaging events (windfall, bark beetle infestations) to forests and iv) climate protection itself (e.g. indirect land-use effects). These interdependencies can also have a major effect on the GHG emissions reduction potential of a particular measure. This is particularly relevant for the potential GHG reductions associated with the substitution of bioenergy for fossil energy which can be substantially affected by systemic changes in land use (e.g., land-use change resulting from an expansion of cultivated area).

Considering all relevant feedbacks is a major scientific challenge, yet crucial for the development of robust strategies to deal with climate change.

Agriculture can reduce GHG emissions and strengthen carbon sinks in many ways. For any given level of agricultural production, the largest potentials are related to ruminant feed, fertilization practices, reduction of nitrogen losses, and increase of nitrogen efficiency (very likely). Sustainable strategies to reduce GHG emissions in agriculture require resource-conserving and efficient cultivation concepts including ecological agriculture, precision agriculture, and plant breeding that preserves genetic diversity.

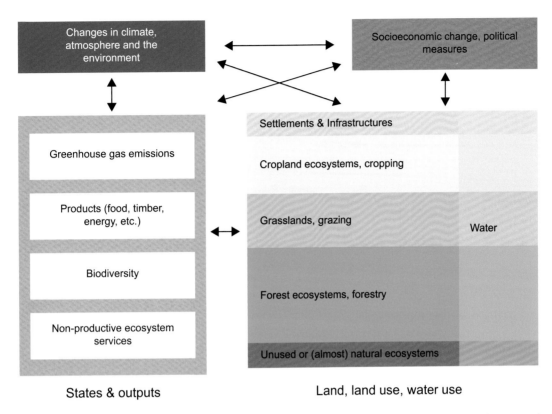

Figure S.3.3. Land systems are characterized by intensive systemic feedbacks between different components such as the society, the economy, climate and climate change, ecosystems, etc. Activities to reduce GHG emissions or to adapt to climate change therefore often cause numerous additional effectsimpacts. Source: Adapted from GLP (2005); MEA (2005); Turner et al. (2007)

Between 1990 and 2010, climate-relevant emissions from the agricultural sector fell by 12.9 % in Austria. This was due initially to a reduction in the number of animals (until 2005) and then (2008 to 2010) to a reduction in the use of nitrogen fertilizer. During the same time period the number of pigs and cattle increased again, which led to an increase in emissions from ruminant digestion and manure. With emissions at 7.5 Mt CO_2-eq. in 2010, agriculture was responsible for 8.8 % of Austria's total GHG emissions (Volume 3, Chapter 2).

Increased production of agricultural bioenergy can contribute to GHG reductions if implemented as part of a strategy for the integrated optimization of food and energy production as well as through "cascade utilization" of biomass. (This strategy proposes optimizing the integrated use of biomass as a raw material and as an energy carrier.) The potential of agricultural areas to reduce GHGs can be increased through an integrated optimization of crop rotation, animal husbandry, and biomass utilization for food, fiber, and energy production. At the same time, energy and water balances and biodiversity preservation need to be considered systematically (Volume 3, Chapter 2).

Adaptation measures in the agricultural sector can be implemented at varying rates. Within a few years measures such as improved evapotranspiration control on crop land (e. g., efficient mulch cover, reduced tillage, wind protection), more efficient irrigation methods, cultivation of drought or heat-resistant species or varieties, heat protection in animal husbandry, a change in cultivation and processing periods, as well as crop rotation, frost protection, hail protection, and risk insurance are all feasible (Volume 3, Chapter 2).

In the medium term, feasible adaptation measures include soil and erosion protection, humus build up in the soil, soil conservation practices, water retention strategies, improvement of irrigation infrastructure and equipment, warning, monitoring, and forecasting systems for weather-related risks, breeding stress-resistant varieties, risk distribution through diversification, increase in storage capacity as well as animal breeding and adjustments to stable equipment and to farming technology (Volume 3, Chapter 2).

In principle, adaptation measures in the agricultural sector can be decided upon or mandated at the level of the farm or higher levels (private/public forums); implementation, however, always needs to take place at farm level. Adaptation measures can be more or less autonomous, for instance, when climate change influences the phenology of plants – that is,

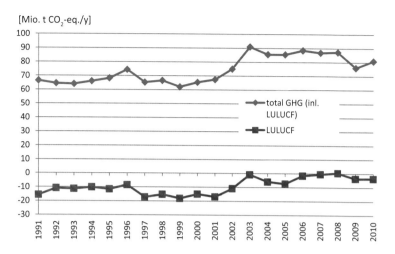

[Mio. t CO_2-eq./y]

Figure S.3.4. Total Austrian GHG emissions (including sources and sinks from land use, land-use change and forestry, LULUCF) contrasted with LULUCF emissions only. Source: National Inventory Report, Anderl et al. (2012)

when there are temporal changes in the annual cycle – that will affect production-related measures. Adaptation measures can also be the result of a conscious choice (planned) among various options, for example, changing crop rotations, type of cultivation, or soil management. From a societal perspective it makes sense to consider not only the economic "benefits" and "costs" of adaptation measures, but to evaluate them in terms of their contribution to sustainable land management and GHG reductions (Volume 3, Chapter 2).

The type of forest management and the rate of wood use have a large influence on the carbon cycle. Until 2003 Austrian forests were a significant net CO_2 sink; since then, this function has decreased and in some years has been close to zero. The sink function of the forests up to 2003 was due to both growth in forest area and an increase in stored carbon per area unit. With the exception of the past ten years, carbon stores have increased significantly over the past decades as felling has been continuously lower than growth. More recent decreases in the carbon sink can be explained by the significant increase in harvests after 2002 and a series of large-scale disturbance events (storms, bark beetles). Furthermore the calculation method was changed: for the first time changes in soil carbon (litter layer and mineral soil) were incorporated, showing soils to be a slight carbon source (Volume 3, Chapter 2).

The GHG emission balance of forest biomass depends heavily on systemic effects in the forest sector. Interactions must be considered between the amount of wood harvested, the carbon sink function of the forest, and the accumulated carbon stock that, depending on thwe period under consideration, yields different net GHG emissions (Figure S.3.4, Volume 3, Chapter 2).

Replacing emission-intensive resources or construction elements in long-living products, particularly buildings, with

wood can contribute to increased carbon storage in products and to total GHG reductions. With regard to the GHG balance, the best results are yielded by an integrated optimization of forestry, including forest management, use of wood for long-living products, and use of by-products for energy, such as wood residues and wood waste from production and products at the end of their lifetime. In many cases, cascading use of biomass in forestry is an ecologically effective use strategy. Depending on the location, type of tree species, economic conditions etc., the use of wood residues (i. e., industrial roundwood, harvest residues) as fuel can make sense. A utilization plan of this kind is in effect in large parts of Austria (Volume 3, Chapter 2).

As forestry requires long-term planning, adaptation to climate change is a particular challenge. Despite considerable uncertainties, decisions need to be taken today that are appropriate for the new climate conditions. An "appropriate" strategy in forest management would be one that provides the sectors agents with sufficient leeway for action to deal with unexpected developments.

Particular challenges would be uncertainties pertaining to the regional distribution of changes, particularly for extreme weather events, and the risk of insect pests and fungi that are harmful to forests.

The challenges of climate change for forestry vary greatly from region to region. In regions where the productivity of forests is currently limited by the length of the vegetation period, climate change will increase productivity. This is true for large parts of mountain forests and areas that are above the current tree line. Regions, mainly in south-eastern and eastern parts of Austria, that are today subject to drought-related problems and related insect damage will become more difficult to manage in future (Volume 3, Chapter 2). The most sensitive areas are Norwegian spruce stands located in lowlands and pure spruce forests which serve a protective function in mountain regions. The adaptation measures in the forest sector are associated with considerable lead times (Volume 3, Chapter 2).

The resilience of forests to risk factors as well as the adaptability of forests can be enhanced. The choice of tree species and rotation period is an important parameter for adaptation strategies particularly to reduce the risk of damaging events, although interactions with the carbon sink function of forests need to be considered (Volume 3, Chapter 2). Furthermore, small-scale silviculture leading to heterogeneous forest structures is considered to be a suitable adaptation means. A survey has shown that managers of forest enterprises are already aware of the relevance of climate change for forestry. Over 85 % of managers of large forest enterprises have indicated that they have already implemented climate change adaptation measures. In contrast, owners of small forests have not reacted much to date (Volume 3, Chapter 2).

Successful **adaptation of water management to climate change** can be ensured by means of integrative, interdisciplinary approaches. Adaptation measures in the areas of floods and low water, such as land-use change in the watershed, can contribute to GHG reductions by carbon sequestration. Changes to the solid material budget through rising global temperatures have less negative impacts on flowing waters than the absence of the sediment continuum. For the provision of drinking water, important adaptation measures would involve linking small suppliers and amenities and the creation of redundancies of virgin water sources. The main challenge for wastewater treatment is to account for decreased water flows in the receiving waters. Increasing organic content in soil leads to increased storage capacity for groundwater. Through the protection and expansion of water retention areas (e.g., floodplains), objectives of flood and biodiversity protection to adapt to changing discharge conditions can be combined (Volume 3, Chapter 2).

In water management there are only few possibilities for reducing GHGs. In urban water management the construction of suitably large digesters for the production of biogas in water treatment plants can contribute to GHG reductions. It is difficult to avoid methane emissions from existing reservoirs (Volume 1, Chapter 1).

Studies have come to differing conclusions regarding the impact of climate change on the energy production of hydroelectric plants. However, production is expected to shift from summer to winter (Volume 3, Chapter 2).

Climate change increases pressure on ecosystems and biodiversity, which are currently already burdened by numerous factors such as land use and emissions. Many nature pro-

tection measures that promote biodiversity can also contribute to GHG reductions. Protection and restoration of moors or decreasing the intensity of use of key forests or wetlands creates carbon sinks and promotes biodiversity. Such measures can be economically attractive, but will not be implemented to any significant extent without incentives (Volume 3, Chapter 2).

Ecosystems and biological diversity are threatened not only by climate change but also by many other global, regional, and local changes. The introduction of foreign invasive species, deposition of toxic substances, destruction of habitats due to housing construction, trade, industry, or tourism, water use, and agricultural and forestry changes, for instance, can all have negative impacts. Measures in other sectors have both indirect impacts (via climate change) and direct impacts, such as land use, on nature protection, ecosystems, and biodiversity. GHG reduction measures in other sectors are often also adaptation measures for nature protection and biodiversity (Volume 2, Chapter 3; Volume 3, Chapter 2).

Increasing pressure on ecosystems and biodiversity can lead to a loss in the capacity of ecosystems to deliver an adequate quantity and quality of critical ecosystem services. In particular, risks arise from ecosystem deficiencies that are already present and through climate-induced shifts of habitat boundaries where species are unable to cope with due to migration barriers, for example, in the alpine region. Creating a comprehensive habitat network in Austria is an important adaptation option (Volume 2, Chapter 3; Volume 3, Chapter 2).

Trade-offs between climate protection measures and biodiversity protection can occur. **In the area of renewable energy, for instance, conflicts between climate protection and biodiversity arise.** Further expansion of hydropower can lead to a decrease in biological diversity in flowing waters. Increasing land use for the cultivation of energy crops or intensive use of forests can impair their functions as carbon sinks and have impacts on biodiversity. Early identification of possible conflicts between climate and biodiversity protection enables the best use to be made of existing synergy potentials (Volume 3, Chapter 2).

Sustainable consumption offers significant GHG reduction potentials. Demand-side changes, such as changes in dietary consumption habits and measures to reduce food waste, can make a significant contribution to GHG reductions (Volume 3, Chapter 2).

In the EU25, almost 30 % of total GHG emissions from consumption can be attributed to food. The consumption of meat and dairy products causes 14 % of total GHG emissions in the EU27. In Austria, GHG emissions of food consumption are likely to be similar to those in Germany, where roughly half of GHG emissions related to food consumption come from agricultural production – 47 % from meat and 9 % from plant production. The remaining 44 % of GHG emissions from food consumption come from processing, trade, and consumption activities such as cooling etc. (Volume 3, Chapter 2).

Significantly reducing animal products in the diet can contribute considerably to GHG emission reductions. A regional, seasonal, and predominantly vegetarian diet as well as the consumption of products with lower GHG emissions across the entire supply chain can result in significant GHG savings. A greater consumption of organic products, combined with a change in consumption toward more plant products to compensate for the lower output of organically grown products and thus the extra growing space required, can also contribute to GHG reductions. Overall, it is estimated that an extensive change in diets could save over half of GHG emissions related to food production (Volume 3, Chapter 2). Such behavioral changes also have significant knock-on effects in terms of health improvements (Volume 2, Chapter 6).

Decreasing losses in the entire life-cycle (production and consumption) of food could make an important contribution to GHG reductions. However, data on food loss and waste in Austria are contradictory and not very robust; in some cases the avoidance potential is quite low, when compared internationally. Substantially more research is needed (Volume 3, Chapter 2).

Systemic effects cause large uncertainties in the comprehensive assessment of GHG effects of bioenergy. These effects relate particularly to direct and indirect effects of land-use change. Land-use related GHG emissions from bioenergy production can be either positive or negative. In many cases they are the decisive factor in determining whether replacing fossil energy by bioenergy actually achieves the desired GHG emission reduction effect. The extent of GHG emissions related to land use change depends mainly on two factors: i) the usage history of the land to be deployed for bioenergy cultivation and the characteristics of the bioenergy crops and ii) systemic effects such as the displacement of cultivation of feed- and foodstuffs (indirect land-use change: ILUC). The uncertainties regarding data and models for estimating ILUC are no justification for ignoring the systemic effects related to large-scale cultivation of bioenergy. Ignoring these would be an implicit assumption that emissions related to ILUC are zero, which in general is incorrect. Emissions related to ILUC must be included in calculations of GHG emissions to determine if the cultivation of bioenergy crops contributes to GHG reductions (Volume 3, Chapter 2).

S.3.3 Energy

Energy is vital for our economic system and for the production of goods and the delivery of services: energy use thus offers a particular large range of possibilities for taking action and shaping transformation. While energy use per value added (energy intensity of GDP) has barely changed since 1990 in Austria, it has decreased significantly in other countries and at the EU average. From 1990 to 2011 between 4.8 and 5.5 PJ of primary energy were used per billion EUR of gross domestic product, without any relevant change in trend. From a climate perspective, however, this is a significant problem as the transformation of primary *fossil* energy into energy services is accompanied by GHG emissions.

For a transformation of the energy system, a focus on energy services is crucial. Energy services are the actually relevant factor for prosperity. Energy productivity at all levels of the energy system defines the amount of energy that is necessary for the delivery of the energy service. The energy mix defines which energy source is used directly for end use or for input to transformation processes. Climate policy thus needs to address all three areas of energy demand, energy technologies, and type of energy sources (Volume 3, Chapter 3; Volume 3, Chapter 6).

In 2011 total gross domestic energy demand for primary energy in Austria was over 1 400 PJ, with an energy input that has more than tripled since 1955 (Figure S.3.5). At the same time primary energy demand stagnated between 2005 and 2011 with a significant reduction in 2009, which can be attributed to lower levels of production during the economic crisis (Volume 3, Chapter 3).

During this time, fossil fuels have dominated with a share consistently over 70 %, which in absolute figures equates to an increase from approximately 750 PJ (1973) to approximately 1 000 PJ (2011). Within the fossil energy mix, the share of coal has reduced significantly both proportionally and in absolute terms (from 245 PJ to 145 PJ; Volume 3, Chapter 3).

Demand for petroleum products has increased from 450 PJ to 550 PJ since 1973, an increase which is due exclusively to the transport sector. In other sectors (industry, electricity production, heating) the use of oil has declined considerably. Gas is the only fossil energy source whose share of primary energy consumption has increased (in 2011, at around 350 PJ, the share was almost 24 %). The share of renewable energy had risen to 26 % by 2011. Historically, the most important aspects in the development of end energy use were the increases in the shares of electricity (from 17 % in 1990 to 23 % in 2011) and of gas (from 13 % in 1990 to 28 % in 2011; Volume 3, Chapter 3).

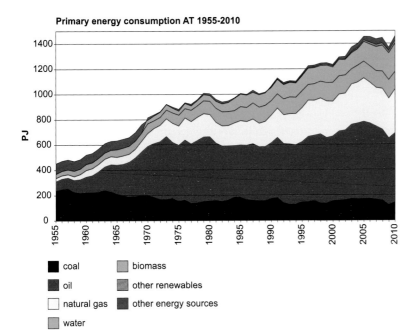

Primary energy consumption AT 1955-2010

Legend:
- coal
- oil
- natural gas
- water
- biomass
- other renewables
- other energy sources

Figure S.3.5. Primary energy consumption in Austria by energy sources. Source: Graph by R. Haas based on data of the Energy Economics Group and Statistik Austria (2013)

Between 1990 and 2011 the share of energy-related GHG emissions was approximately 87% of total emissions. Energy-related GHG emissions depend on the demand for energy services, the efficiency of the transformation technology, and the specific GHG emission factor. The main reasons for the high GHG intensity of the Austrian energy system are the high transformation losses (around 50%) from primary to useful energy, the high share of GHG emitting fossil fuels (currently around 70% of Austrian energy use), and also the low energy prices, which on average have not changed when accounted for in real prices (i. e., adjusted for purchasing power). Since 1990 energy-related GHG emissions in Austria have risen practically only in the transport sector (up to almost 25 Mt CO_2-eq. by 2005, slightly declining thereafter). Conversely, the household sector has registered a decrease of around 20% since 1990. All other sectors have only seen very marginal change (Volume 3, Chapter 3).

Due to its high share of GHG emissions and the numerous mitigation possibilities, the energy sector is very relevant **for climate protection.** The energy sector also offers a number of synergistic measures that achieve both GHG reductions and adaptation effects (e. g., passive measures to reduce the cooling load of buildings, photovoltaics as an additional supply capacity element in summer).

The most important options for mitigating GHG emissions in each of the individual sections of the energy supply chain are the following:

Energy production: in principle, GHG emissions in primary energy use can be reduced by the use of renewable energy sources, and also by carbon capture and storage (CCS) technologies or the use of nuclear energy.

In Austria the latter two are not considered to be possible options. Accordingly, only the use of renewable energy is discussed here. The potential of all available renewable energy sources in Austria by 2050 is approximately 170 TWh or 610 PJ per year, of which biomass, wind and photovoltaics could particularly provide a significantly higher contribution than they currently do (Volume 3, Chapter 3).

Energy transformation and transmission: Depending on the scenario, up to 100% of electricity generation can be covered by renewable energy technologies by 2050. The recent market entry of renewable energy represents the most significant change in electricity production. Due to currently continuously declining costs, particularly in photovoltaics, the significant increase in renewables is set to continue. In the next years this will change the entire Austrian market system, as very large amounts of electricity are temporarily produced by these plants, increasing the share of own-consumption; at the same time electricity storage and smart grids will play a considerably more important role in the electricity system then they currently do (Volume 3, Chapter 3).

To optimize these developments in the energy system, changes in infrastructure and structural adjustments to production, networks, and storage will be necessary. Barring possible socio-political concerns regarding, for example, data and privacy protection changes and adjustments are absolutely achievable as long as i) energy political framework conditions are developed and decentralized renewable energy technologies are deployed in production accordingly, ii) smart grids are implemented at the level of distribution systems, iii) new electricity storage technologies and capacities are established and iv) smart meters are installed at the user level (Volume 3, Chapter 3).

Changes have already been registered in the area of district heating (buildings have very low heat density following thermal renovation) for centralized district heating networks; at the same time, district heating has the potential to facilitate the transition to a renewable heat supply (Volume 3, Chapter 3; Volume 3, Chapter 5).

Energy use: On the demand side, options to reduce energy demand are high quality thermal renovation of buildings to heat and cool residential housing and increased, as well as optimized integration of renewables. Current developments of increasingly ambitious new building standards can make a valuable contribution to climate protection and energy efficiency. Under these circumstances, about 70 % of the energy demand for significantly thermally improved buildings by 2050 can be covered from renewables, for which a broad portfolio of biomass, solar heat, and geothermal energy could be deployed (Volume 3, Chapter 3; Volume 3, Chapter 5).

In Austria there is significant energy saving potential in electricity use, although studies clearly show that demand will continue to rise considerably without significant political interventions and unless there is a portfolio of effective measures (Volume 3, Chapter 3).

Options to adapt to climate change: The need for the energy sector to adapt to climate change relates particularly to the climate-dependency of renewables, increased cooling demand of thermal power plants, and changes in energy demand through a shift in heating and cooling needs. Potential impacts of climate change are particularly significant for hydropower in Austria, due to changes in precipitation amounts and patterns (especially seasonal shifts) and changes in runoff through increased evaporation. This can also affect conventional thermal power plants indirectly via the availability of cooling water. These developments could be countered by changes in turbines or in reservoirs, to either secure or even increase energy output.

Energy policy instruments: Implementing mitigation measures will require energy policy instruments, which can be summarized from the studies and scenarios that were analyzed in the following portfolio (Volume 3, Chapter 3).

- *CO_2-related energy tax:* The central instrument of most policy studies is the implementation of continually increasing energy taxes to effectively reduce GHG emis-

sions, combined with incentives to change to less CO_2-intensive energy sources and to increase energy efficiency; this means decreasing energy use but also increasing investments in energy-efficient appliances, vehicles, and facilities. Competition on the energy market and the related switch to cheaper (renewable) energy means that the initially significant subsidies are no longer given for, for example, biological energy sources, hydrogen, or electricity for e-mobility. The lower taxation rates for these fuel types shift demand, which results in environmental benefits in the long-term.

- *Standards:* Dynamic maximum consumption standards are particularly important instruments in various areas: i) tightening thermal building standards for existing buildings, ii) implementing thermal standards for new buildings (analogous to plus-energy-houses), iii) tightened standards for electric appliances in households and in the service industry (office buildings), and iv) rigorous tightening of standards for CO_2 emissions from various alternative energy sources, are all important standards for reducing and optimizing energy consumption.

- *Other incentive systems:* Subsidies are expedient in areas where the preferred means of financial support for renewables are feed-in tariffs or market premiums, particularly as long as there are no taxes that incorporate all externalities; in addition, incentives for the increasing market integration of renewables into electricity production and also for heating and mobility are helpful, as are subsidies for ecological building renovation and explicit incentive and information systems (e. g., labelling systems) for the elimination of old and unviable appliances.

- *Soft knowledge and skills:* Increasing the general knowledge level regarding energy saving is necessary to combat energy poverty and for targeted appliance exchange and renovation activities. Improved advice for exchanging heating systems, electric appliances, and renovating buildings are also part of soft knowledge and skills. In the housing sector especially, there is a great need for auditing and monitoring activities to successively identify energetic weaknesses.

Conclusion. The following basic incentives exist to decrease GHG emissions in the energy sector:

- Reducing demand for energy services, for example, heating / cooling, electric appliances, motor vehicle use.
- Improving the efficiency of the energy supply chain, namely, more efficient provision of the energy service, for

example, more efficient electric appliances, and lower fuel intensity of motor vehicles at unchanged performance and service level.

- Providing the entire energy services demand with energy sources with low CO_2 emissions, for instance, through a shift to renewable energy sources.

Studies with ambitious energy or GHG emission reduction scenarios assume that on average the use of renewables can be increased by up to at least 600 PJ by 2050. If it were possible to lower total energy use to the available level of renewables in the same time frame, a GHG-free energy supply would be possible by 2050 (Volume 3, Chapter 3).

In conclusion, the following can be noted: only if a co-ordinated mix of these individual measures is implemented and prevailing conditions in society are considered, will it be possible to broadly unlock the GHG reduction potential in Austria by 2050 (Volume 3, Chapter 3; Volume 3, Chapter 6).

S.3.4   Transport

Of all sectors, greenhouse gas emissions increased most in the transport sector, by 55 %, in the last two decades. The regulatory instruments used in the EU in the past years – essentially standards for CO_2 emissions per km – failed to take effect for a long time because most of the increased efficiency of vehicles was compensated for by increased driving distances and larger / heavier vehicles (Volume 3, Chapter 3).

Assuming that increasing transport performance and ve-hicle kilometers travelled will be as indicated in the Austrian transport forecast 2025+ with no additional measures, and taking into account the technical regulations (EU and national level) that have already been decided, CO_2 emissions in the transport sector can be expected to continue to rise over the next few years. The technical thresholds that have been agreed will lead to a decrease in CO_2 emissions only by the middle of this decade; emissions would still be 12 % above 1990 val-ues in the year 2030. In 2030 around 45 % of transport- re-lated CO_2 emissions would originate from passenger cars and roughly 35 % from road-based goods transport (values do not consider air traffic; Volume 3, Chapter 3).

The first successes from the limitation of CO_2 emissions per kilometer for passenger cars and delivery vans are already visible. In the past, changes in public transport supply and (perceptible) price signals have also had an effect on the share of private motorized transport in Austria (Figure S.3.6; Vol-ume 3, Chapter 3).

To achieve a significant reduction in greenhouse gas emissions from passenger transport, a comprehensive package of measures is necessary. Key to achieving this are marked reductions in the use of fossil-fuel energy sources, increasing energy efficiency, and changing user behavior. A prerequisite is improved spatial structures of economic pro-duction and settlement, where the distances that need to be travelled are minimized. This could strengthen the environ-mentally friendly forms of mobility used, such as walking and cycling. Public transportation systems could be expanded and improved, and their CO_2 emissions minimized. Technical measures for car transport include further massive improve-ments in vehicle efficiency of vehicles or the use of alternative power sources – provided that the necessary energy is also pro-duced with low emissions (Volume 3, Chapter 3).

Freight transportation in Austria, measured in ton-kilometers, in the last decades increased more than gross domestic product. The further development of transport de-mand could be shaped by a number of economic and social conditions. Optimizing logistics and strengthening the CO_2 efficiency of transport are two potential control functions. A reduction in greenhouse gas emissions per ton-kilometer can be achieved by alternative power and fuels, efficiency improve-ments, and a shift to rail transportation (Volume 3, Chapter 3).

Substantial reductions in GHG transport emissions re-quire a coordinated portfolio of political measures, which include avoiding transport (reduction of distance travelled), a shift to more efficient transport modes (public transport), and the use of "zero-emission" vehicles and renewable ener-gy (Figure S.3.7). Central aspects are appropriate economic framework conditions, namely, new price (for motorized pri-vate transport) and tariff systems (for public transport) as in-centives to shift from motorized private transport to public transport and zero-emission vehicles (Volume 3, Chapter 3).

Spatial planning measures could also contribute to a reduc-tion in vehicle kilometers (passenger and freight) travelled, by placing people's basic needs (living, working, education, recre-ation, community) and economic exchange processes close to-gether. Higher efficiency in car transport can also be achieved by higher occupancy rates (carpools, fewer unladen journeys, less searching for parking spaces etc.; Volume 3, Chapter 3).

Reducing the use of fossil energy requires combustion mo-tors with lower consumption or measures that promote ve-hicles with lower CO_2 emissions (such as, for example, e-mo-bility, generated by renewables) and an increase in the energy efficiency of traffic flows (Volume 3, Chapter 3).

Spatial planning: From a spatial planning perspective, the biggest adaptation successes are to be expected in the Alpine

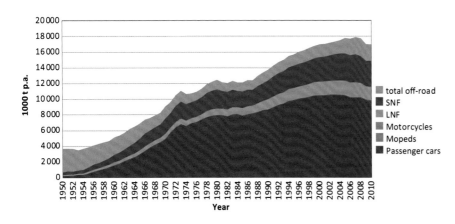

Figure S.3.6. Historical development of CO_2-emissions in transport from 1950 to 2010 in Austria; LNF = light commercial vehicles (<3.5 t total weight); SNF = heavy duty vehicles (>3.5 t total weight and buses); Off-road = trains (steam and diesel traction, construction machines, agricultural machines, lawnmowers, etc.). Source: Hausberger and Schwingshackl (2011)

region through the general further development of planning instruments and cooperation across countries and sectors. The resulting knowledge transfer and the accompanying awareness raising should offer avenues for developing robust settlements and infrastructures, as well as protection against natural hazards, optimal management of water and other resources; these sustainably provide landscape development and safeguard open spaces and also bring about a reorientation of tourism (Volume 3, Chapter 3).

Economic sphere (taxes and subsidies): All studies analyzed with regard to new pricing systems for motorized private transport, identify similar priorities: continually rising (CO_2-based) fuel taxes could be particularly effective, supplemented by a consumption-based vehicle registration tax, which should avoid trends toward larger vehicles and ensure greater efficiency. Supportive measures could include road-pricing in larger cities, abolishing benefits for company cars, revenue-neutral reshaping of commuter tax breaks, development of new concepts for and intensification of parking management and simplifying tariffs for public transport together with increasing incentives to purchase season tickets (Volume 3, Chapter 3).

The significant effects of price increases on energy- and GHG intensive forms of mobility in favor of reduced kilometers driven and / or a shift to other means of transport or public transport have been scientifically validated (Volume 3, Chapter 3).

Traffic planning and "soft tools": Inducing shifts in passenger transport requires the further development of public transport and increased incentives for its use, better mobility management in companies, supporting bicycle transport (building new cycle paths, closing gaps in existing cycle path networks, building bicycle parking spaces) and convincing public relations work (Volume 3, Chapter 3; Volume 3, Chapter 6).

Implementing a shift in freight transport requires improved logistics, higher capacity utilization of haulage (in terms of

weight and volume), and increasing the attractiveness of rail and inland water (Danube) vessels by developing rail routes and connections to shipping infrastructure (Volume 3, Chapter 3).

Technological solutions for alternative drive technologies, alternative energy forms, and increased efficiency in conventional "vehicles": More efficient technologies include primarily the increased use of alternative fuels and an increased share of electrically driven passenger cars and light commercial vehicles and the reduction of specific CO_2 emissions of biofuels, so that they emit 70 % less than fossil fuels by 2020. The extent of reductions is limited by the emissions that occur during the production of biofuels, so that their large-scale use is increasingly being called into question (Volume 3, Chapter 2; Volume 3, Chapter 3).

Until 2030 the relevance of alternative fuels (biofuels, hydrogen, and natural gas) will remain within moderate limits. Electrification of road freight transport is currently not sensible, leaving biofuels as the only viable alternative both in these applications and for mobile machinery (Volume 3, Chapter 3).

S.3.5 Health

The Austrian health system can make a significant contribution to an equitable climate transformation. The Austrian health and welfare systems employ approximately 10 % of the workforce and produce approximately 6 % of Austrian gross value added; this share is growing. This important role in the Austrian economy also entails a large responsibility of the sector for the sustainable development of the services it supplies. As ecological sustainability is important for the long-term support and preservation of health, the health sector has a role model function, which underlines its responsibility in the context of climate protection (Volume 3, Chapter 4).

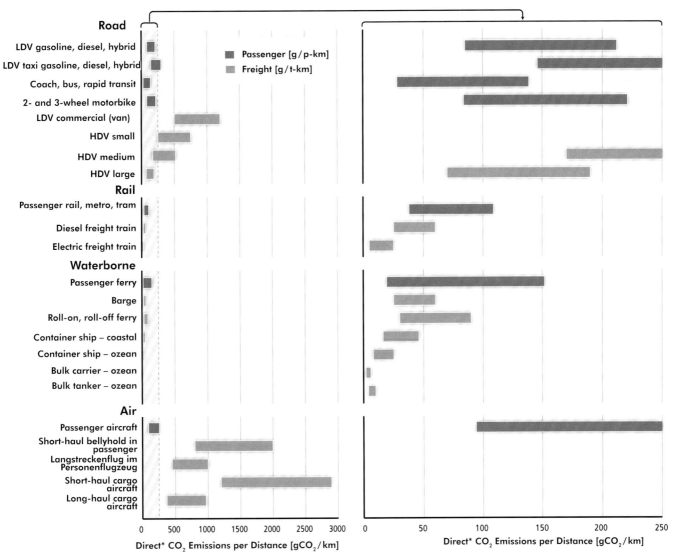

Figure S.3.7. A comparison of characteristic CO_2-emissions per passenger-kilometer and ton-kilometer for different transport modes that use fossil energy and thermal electricity generation in the case of electric railways. Source: IPCC (2014)

Many measures in health care were not specifically developed for the sector, but are part of sectoral strategies. Measures such as high thermal building standards, efficient energy management, and the transition to renewable energy sources also possess high potential for emission reductions in the health sector (see strategies in the section on "buildings" and Volume 3, Chapter 5), as a number of pioneers have already demonstrated. Health care has particular opportunities to reduce emissions in the areas of mobility, environmentally and resource-friendly procurement and climate-friendly waste man-

agement. In this regard, establishing incentives for patients and staff to engage in climate-friendly behavior can make a significant contribution (Volume 3, Chapter 4).

Adaptation in the context of health relates to institutionally and privately planned measures on the one hand and to biological-physiological processes on the other. The latter are automatic, subconscious processes in human bodies and take place at different levels and at different speeds. In this context it is important to gain knowledge about the existing borders of such adaptation processes and the high-risk groups who, due to various factors (age, medical history, social factors etc.), possess limited adaptation abilities. While biological-physiological processes tend to allow adjustments to be made to short-term weather events, institutional adaptation strategies help adjustment to long-term changes and support of the high-risk groups (Volume 2, Chapter 6; Volume 3, Chapter 4).

The health system is thus a central element in improving the capacity to adapt to the possible health impacts of climate change. High-risk groups should be particularly supported, as they are sensitive to climatic changes due to age or medical history. Furthermore, sustainable health care focuses on prevention rather than on the treatment and cure of illnesses. Such a transformation requires structural changes to be made to the entire system (Volume 3, Chapter 4).

Risks, as in newly introduced or established pathogens and vectors are almost impossible to predict and the possibility of taking prophylactic action is very small. They are therefore a big challenge for the health system (Volume 2, Chapter 6). Continual and detailed collection and monitoring of health data, which need to be regularly linked to climate and proliferation data, serve as an important basis for developing targeted adaptation strategies. To date, such studies are temporally selective or geographically limited to a few regions in Austria (Volume 3, Chapter 4).

A barrier in the health care context is the limited availability of data. Although the health system routinely collects health data, these are either not available or not available in sufficient detail for scientific research. Sufficient data as a basis for conceptualizing adaptation strategies and without which meaningful and detailed analyses of regional and local dose-effect-relationships, are hard to generate; this area is currently hampered by concerns over data protection, unclear competences, lack of cooperation, and technical problems (Volume 3, Chapter 4).

In any case, health-related adaptation can entail behavioral changes on the part of many individuals, of large parts of the population, and of members of particular high-risk groups (Volume 3, Chapter 4).

Finally it is important to note that adaptation and mitigation measures in other areas can also be relevant for human health. It is thus important to avoid negative feedbacks and, conversely, to make use of synergetic effects between different areas and sectors (Volume 3, Chapter 4).

Climate-relevant transformation is often directly related to health improvements and accompanied by an increase in the quality of life. Shifting from car to bike use, for example, has a proven positive preventive impact on cardiovascular diseases and has other health-improving effects that significantly increase life expectancy, in addition to positive environmental impacts. Sustainable diets have also proved to have health-supporting effects (e.g., reduced meat consumption). Due to existing feedback effects, total effectiveness is raised when health experts are given a say in the design and planning of relevant measures outside the health system. Only this would make it possible to conceptualize measures so that they are either advantageous to health, or at least that the positive effects outweigh the negative (Volume 3, Chapter 4).

S.3.6   Tourism

Globally, the contribution of tourism is estimated to amount to be around 5 % of total CO_2 emissions; emissions are created due to travel (point of departure to destination), and accommodation and activities (on site). Some 75 % of emissions are caused by tourism due to transportation of tourists, and 21 % by accommodation (Figure S.2.8; Volume 3, Chapter 4).

Tourism, a significant economic sector, can also be assumed to be responsible for a relevant and high share of GHG emissions in Austria. When indirect economic effects were accounted for, tourism contributed 7.45 % to total value added in 2010. To date detailed analyses of emissions from domestic tourism are lacking; detailed data are available only for snow-based winter tourism. There, the largest cause of emissions is accommodation at 58 %, followed by transportation at 38 %. Cable cars, ski lifts, ski slope maintenance vehicles, and snow cannons cause only 4 % of total snow-based winter tourism emissions (Volume 3, Chapter 4).

High savings potential with regard to GHG emissions caused by tourism can be identified in transport and accommodation, and could be achieved by adapting the operational management of tourist facilities.

Successful pioneers in sustainable tourism are showing ways to reduce greenhouse gases in this sector. In Austria there are flagship projects at all levels – individuals, municipalities, and regions – and in different areas, such as hotels, mobility, and tourist activities. Due to the long-term investments involved in infrastructure, tourism is particularly susceptible to lock-in effects (Volume 3, Chapter 4).

Changes in climate have a significant effect on the Austrian tourism industry. This is due to the high dependence on local climatic conditions. On the basis of current knowledge about future climate development, it can be assumed that the consequences will be both negative (in winter) and positive (in summer); but economically the negative effect will be more pronounced because winter tourists spend more. Guaranteeing the long-term and sustainable development of the tourism

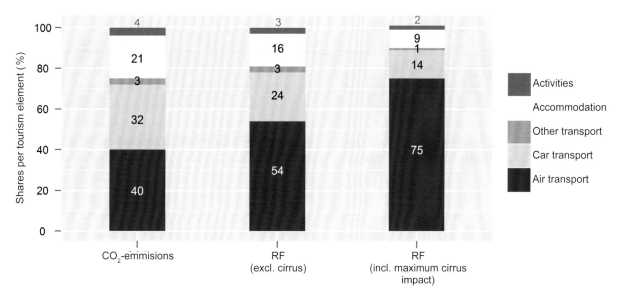

Figure S.3.8. Estimated share of tourist activites which contribute to global CO_2 emissions and radiation (inlcuding day-trippers) in 2005. Source: adapted from UNWTO-UNEP-WMO (2008)

industry depends on timely recognition of the advantages and benefits of climate change as well as an adaptive strategy based on these insights.

Various areas of Austrian tourism will be affected by climate change in different ways. It is expected, for instance, that **city tourism will experience barely any net annual change, but considerable seasonal shifts**. It is possible that city tourism will decrease in summer due to an increase in hot days and tropical nights. Shifts in tourism flows to other seasons and regions are possible and can already be observed in some cases. For alpine swimming lakes, climate change could even turn out to be advantageous. However, particularly negative effects are to be expected for Lake Neusiedel – whose water level is expected to decline considerably – mountain tourism, and alpine winter tourism. The main problems for mountain tourism, already in evidence today, are the decline in permafrost and glaciers, which implies unstable paths and risks of rock fall. In addition to modifying, maintaining, or indeed creating new paths to alpine huts, high altitude paths, and routes to reduce and avoid disproportionate risks, adaptation measure in mountain tourism also include abandoning old paths and installing new ones, and the installation of path-information systems (Volume 3, Chapter 4).

Winter tourism will come under pressure due to the steady rise in temperature. Compared to destinations where natural snow is plentiful, many Austrian ski areas are threatened by the increasing costs of snowmaking. Therefore, adaptation measures relating to alpine winter tourism are of par-

ticular relevance to Austria. This is due on the one hand to the high climate sensitivity of winter tourism (dependence on snow) and to the important position of winter tourism in the domestic tourism industry on the other hand. This important position is also due to the fact that while the numbers of overnight stays in Austria are roughly equal in summer and winter, the income per guest is significantly higher in winter. Already today, compensating for reduced natural snowfall through artificial snowmaking is a widespread measure used to cope with annually variable snow cover (Volume 3, Chapter 4).

Future adaptation possibilities through artificial snowmaking are limited. Although currently 67 % of the slope surfaces are equipped with snowmaking machines, the use of these is limited by the rising temperatures and the (limited) availability of water (likely, Volume 3, Chapter 4). The promotion of the development of artificial snow by the public sector could therefore lead to maladaptation and counterproductive lock-in effects (Volume 3, Chapter 4).

Snowmaking leads to higher energy use, which in turn leads to higher costs and higher prices for skiers. For many people this is a reason not to go skiing. Another strategy is to expand or change the location of ski resorts to higher altitudes and to north faces to secure continuous operation with an earlier start and later end to the season. Such measures have already been observed in the past. However, this strategy also has several disadvantages, such as i) conflicting with skiers' preference for sunny slopes, ii) the natural landscape limitations of many ski resorts to expand to higher altitudes, iii) increased risks of avalanches, and iv) exposure to wind and the danger to fragile ecosystems (Volume 3, Chapter 4).

A general and often mentioned strategy to adapt to climate change – not just for winter tourism – is **diversifying supply**. Due to the implicit insurance effect, a mixed supply portfolio is subject to lower risk than a one-sided supply. However, results show that the potential for diversifying supply is limited, as ski destinations are visited for snow-based activities not because of activities that can take place independently of snow (Volume 3, Chapter 4).

A strategy of last resort for particularly endangered areas could be the compilation of an integrative exit scenario from snow tourism. Particularly at the edge of the Alps and at low altitudes, some resorts that are no longer profitable have already started to close. A well-known and successful example of an actively planned exit from winter tourism following a number of winters with little snow at the start of the 1990s is the ski resort Gschwender Horn in Immenstadt (Bavaria). The lifts have been removed and the ski slopes restored. Today, the resort is used for summer (hiking, mountain biking) and winter (snowshoeing, ski touring) tourism (Volume 3, Chapter 4).

Generally there are a number of strategies to enable an adequate adaptation of the tourism sector to climate change (Volume 3, Chapter 4). The success of these approaches depends on whether action takes place individually and reactively or in a linked and anticipatory manner. Only linked and anticipatory activities will avoid counterproductive situations (such as higher resource use through increased snowmaking) and enable long-term, successful development of the Austrian tourism sector (Volume 3, Chapter 4).

Losses in tourism in rural areas have high regional economic follow-up costs. As these job losses frequently cannot be compensated for by other industries, structural change of this type often leads to emigration. Peripheral rural areas already face large challenges through waves of urbanization (Volume 3, Chapter 4).

Tourism could benefit in Austria because of the very high temperatures expected in summer for the Mediterranean. Summer tourism could benefit indirectly, as the expected high temperatures in the Mediterranean region will make the Austrian climate comparatively more attractive (Volume 3, Chapter 4).

S.3.7   Production

From 1970 to 1995 energy input into Austrian industry was fairly stable between 200 and 250 PJ/year, but started to rise

steadily thereafter and crossed the 300 PJ mark in 2005 (Figure S.3.9; Volume 3, Chapter 5).

In the period 1970 to 1995 energy use increased hardly at all, while production value and quantity almost doubled. This was because increases in production were compensated for by increases in efficiency i) as part of general technological development and ii) due to structural change in production. Slumps in 1973 and 1980 can be attributed to energy (price) crises. The share of electric energy has remained almost constant (dotted line in Figure S.3.9) at around 30% in the past 30 years. In the last 1.5 decades the trend changed completely, leading to an increase in energy input by almost 50% to over 300 PJ/year (Volume 3, Chapter 5).

Due to the high share of domestically emitted GHGs, the focus of production has been primarily on mitigation measures (as opposed to adaptation strategies). Decreasing emissions of climate-effective gases from energy input into production can take place though a reduction in energy end use on the one hand, and through a shift to energy sources with lower emissions on the other. Process-related CO_2 emissions can be avoided only through innovations in production or products. Reductions in other GHGs (methane, nitrogen oxide, fluorinated hydrocarbons etc.) can only be achieved through process innovations (Volume 3, Chapter 5).

Although climate protection measures have already been implemented in Austrian industry, there is still enormous untapped emission reduction potential. This relates in particular to efficiency measures and the use of renewable energy. How-

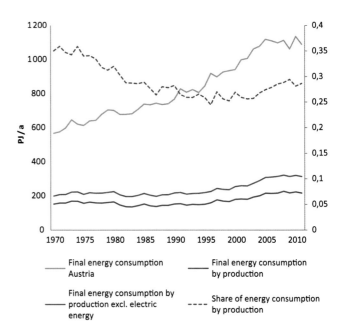

Figure S.3.9. Energy consumption of the production sector in Austria; values in PJ/yr. Source: Statistik Austria (2012)

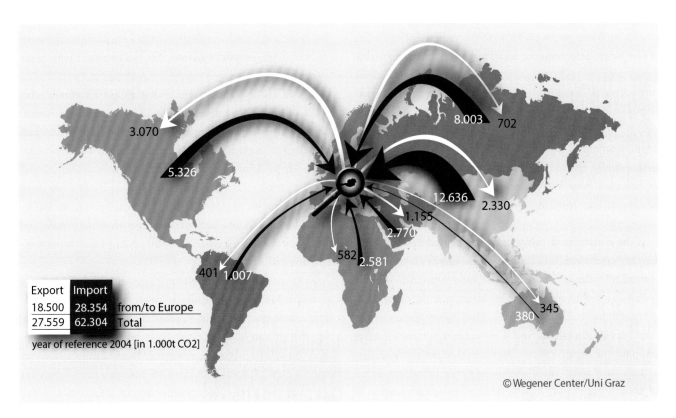

Export | Import |
18.500 | 28.354 | from/to Europe
27.559 | 62.304 | Total

year of reference 2004 [in 1.000t CO2]

© Wegener Center/Uni Graz

Figure S.3.10. CO_2 streams from the trade of goods to/from Austria according to major world regions. The emissions implicitly contained in the imported goods are shown with red arrows, the emissions contained in the exported goods, attributed to Austria, are shown with white arrows. Overall, south Asia and east Asia, particularly China, and Russia, are evident as regions from which Austria imports emission-intensive consumer- and capital- goods. Source: Munoz and Steininger (2010)

ever, emission reductions in line with the 2 °C stabilization target will require the support and development of radical new technological innovations (Volume 3, Chapter 5).

Industry is the largest emitter of GHGs in Austria. In 2010 the share of the production sector in both total Austrian energy end use and GHG emissions was close to 30 %. Emission reductions of 50 % and above cannot be reached through incremental improvements and the application of the state of the art. Such reductions would require either the storage of GHG emissions (carbon capture and storage, as documented in EU scenarios for the 2050 energy roadmap) or the development of new, climate- friendly processes (radically new technologies and products and a drastic reduction of energy end use). This provides an opportunity to develop new materials and products for international markets (Volume 3, Chapter 5).

Only a few subsectors are responsible for a large proportion of energy demand and GHG emissions. The five largest emitting (combustion and process emissions) sectors are iron and steel, metal production, mineral products, pulp / paper / printing, and chemicals. Together, these subsectors are responsible for over two-thirds of total production emissions.

A major emission reduction measure that has already been implemented due to cost benefits is the shift from coal to gas, a very efficient reduction strategy. A downside to this strategy is the resource dependence on countries with insecure and ethically questionable political situations. A number of other voluntary measures have also already been implemented relating to a reduction in fuel demand. In this context, lower fuel demand is often compensated for by higher electricity demand, which improves the emissions balance of the production sector but worsens the emissions balance of the electricity sector. Another measure, to which the largest installations in the energy-intensive sector are subject, is the "EU Emission Trading System." Due to the almost constantly low prices of certificates, emission reduction signals have been rather minor (Volume 3, Chapter 5).

In Austria, while efforts to improve energy efficiency and promote renewable energy can be observed, these lack sufficient measures to reach the targets that have been set. With regard to both energy efficiency and the use of renewable energy sources, the potential is not yet exhausted. With the exception of the pulp industry, the use of renewable energy in industry is not yet prevalent. Depending on the location, small-scale hydropower plants can offer an alternative source of electricity generation. In addition to the use of renewable energy sources, industrial combined heat and power generation is crucial. The paper and pulp industry in particular has very good pre-conditions in this context. There is also significant potential in the generation of electricity from low temperature waste heat (ORC installations). In the medium term some of the technologically necessary carbon can also be procured from biogenic sources. This also requires significant further research (Volume 3, Chapter 5; Volume 3, Chapter 6).

If the emissions in other countries caused by consumption in Austria are included, emission figures in Austria are around 50 % higher. An effective climate protection strategy in industry should consider global processes and include these as a central element. Demand in Austria contributes to the emissions of other countries. If these emissions are included and adjusted for emissions related to Austrian exports, Austrian "consumption-based" emissions can be calculated. These are considerably higher than Austrian emissions as recorded in UN statistics and are on the increase (they were 38 % higher in 1997 and 44 % higher in 2004). The flow of embodied emissions in goods shows that most emissions caused by Austrian imports come from China, southern and eastern Asia, and Russia (Figure S.1.5). Consideration of the global context also puts the sometimes high decreases in industrial end use and the emissions of other EU member states into perspective, as these are often due to a relocation of energy-intensive industry (Volume 3, Chapter 5; Volume 3, Chapter 6).

Currently, none of the sectors investigated have strategies to adapt to climate change. Expected potential challenges are changes in demand for cooling and heating, the availability of bio resources (e.g., wood), and a climate-induced change in demand.

S.3.8 Buildings

Statistics Austria's complete inventory of the number of buildings and homes and the micro census, which contains a statistically relevant sample of homes, form the basis of all Austrian studies on buildings. For commercial buildings, there is an initial study evaluating energy use in various sectors.

The number of commercial buildings and homes in Austria has been increasing on a linear basis since 1961, due to the rising population on the one hand and the increase in space used per person on the other. In 2011 there were around 4.4 million apartments in 2.2 million buildings, of which around 75 % were single and two-family homes. Around 70 % of living space was constructed before 1980 with a low energy standard. A large proportion is suitable for energetic renovation (Volume 3, Chapter 5).

Space heating and other small uses are responsible for 28 % of energy end demand and 14 % of GHG emissions. Despite continued construction of domestic and commercial buildings, energy demand has remained fairly constant since 1996, as additional energy demand by new buildings and energy savings through demolition and renovation are more or less balanced. At 260 PJ / year, domestic households are responsible for roughly 62 % of energy demand and private and public service providers at 130 PJ for about 31 %. The remaining demand is from agriculture. In domestic households, space heating has the main share with over two-thirds (195 PJ / year), domestic hot water is at roughly 13 % (35 PJ / year), and cooking at just under 3 % (7 PJ / year). The rest (remaining 37 PJ / year) is equivalent to domestic electricity demand. Wood, gas, and oil, each provide around 27 % of heating and warm water; district heating provides 14 % and electricity 9 %; solar heat and heat pumps each provide around 2 %.

While the share of renewable energy for domestic households increased from 22.9 % to 26.9 % and district heating increased from 6.9 % to 9.9 % in the period from 2003 to 2010, the share of fuel oil dropped from 25 % to 19 %. Natural gas remained constant at 20.5 %, and the share of coal was very small, all of which demonstrates a clear trend toward renewable energy sources and district heating. This trend is strongly supported by the high volatility of oil prices and the availability of technically advanced automatic heating systems that run on renewables.

The main energy sources in the public and private services sector were electricity (28 %), district heating (23 %), natural gas (20 %) and fuel oil (13 %), while biomass contributes only 2.5 % (renewables were not specified). Pipeline and grid-bound energy sources provide over 80 % of energetic end use in the services sector; at 4.2 %, coal, diesel, petrol and liquid gas as well as renewables and waste played only a marginal role in the sector total.

In 2010 Austrian households emitted 24 Mt CO_2-eq. of GHGs including biomass (equivalent to 26 %). If the CO_2 emissions from biogenic energy sources are calculated as being CO_2-neutral, as is the case internationally, emissions are reduced to 17 Mt and the share declines to 24 %. Heating and other small use as well as warm water and electricity are each responsible for half.

According to the 2011 Austrian climate protection report, the sectors "space heating and other small use" in households (not including electricity and district heating) contributed

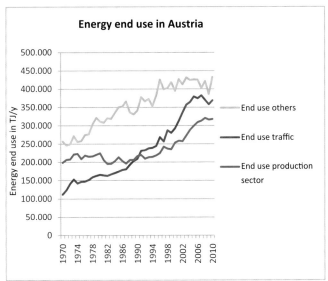
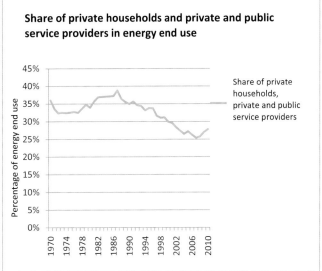

Figure S.3.11. Energy end use according to sector (left) and proportion of households, and private & public sector providers (right). Source: Statistik Austria (2012)

14 % to GHG emissions. This percentage is considerably lower than the share of 28 % of final energy demand because of the use of energy sources that emit less CO_2 (biomass and district heating).

Changes in outside temperature caused by climate change will result in lower heating demand, but will increase buildings' cooling demand. Adaptation strategies in the building sector require legal instruments and subsidies to reduce the cooling demand of buildings and support technical measures relating to the orientation of buildings, window areas, storage space, night-time ventilation, etc.

On the basis of the IPCC IS92a scenario and the algorithms used in the Austrian implementation of the EU *Energy Performance of Buildings Directive* (EPBD), heating demand will reduce by approximately 20 % between 1990 and 2050, while cooling demand increases. However, heating demand will nevertheless dominate cooling demand in most buildings.

Technological progress in recently constructed buildings and renovation has significantly reduced space heating energy demand, which decreased from 42 kWh/m²/year to 28.8 kWh/m²/year in subsidized housing from 2006 to 2010. In accordance with the European buildings directive's (amendment 2010) move toward "nearly zero-energy buildings", implementation of ambitious standards for new buildings is necessary, to achieve long-term climate protection targets. Following thermal-energetic renovation of domestic buildings, the space heating demand reached an average value of 48.8 kWh/m²/year in 2011. In 2006 this value was

at around 67 kWh/m²/year. As the majority of homes are already built, the energetic renovation of buildings is the single most important mitigation measure.

GHG emissions can further be decreased by the optimal use of renewables in buildings. An analysis of the potential of renewables in building use, however, needs to consider the entire energy system including transport, trade, industry and buildings, to avoid an isolated consideration of buildings that yields an over-estimation of potential for this sector.

The lower the energy demand of buildings, the easier it is to supply them with renewables. Solar heat and photovoltaics can increasingly be used for areas that are not necessary for illumination and are oriented accordingly. The scalability to very small sizes and capacity will mean a possible increase in the use of heat pumps operated with renewables. The limited availability of biomass means that its use will be expanded in industry and mobility rather than in buildings, with the exception of self-supply in rural areas. The increasing efficiency of buildings will mean that local district heat networks will play an ever-smaller role, as the ratio of heat provided to network losses becomes increasingly unfavorable.

Without major political interventions, domestic electricity consumption will continue to rise. Although efficient technologies increase the theoretical reduction potential, the increase in electricity intensive applications and constantly low electricity prices implies total electricity consumption to continue to rise, at least moderately.

Through increased efficiency, renewable energy could cover around 90 % of heating demand of buildings by

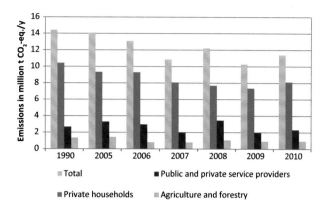

Figure S.3.12. CO_2-equivalent emissions in the space heating sector and other small consumers. Sources: Anderl et al. (2009, 2011, 2012)

2050. The Austrian energy strategy – which for the most part lacks specific measures – allows for an investment of € 2.6 billion/year to achieve a 3 % annual renovation rate for domestic buildings up to 2020. This will trigger gross production value of around 4 billion €/year, with subsidies of around € 1 billion/year. A 3 % renovation rate for commercial buildings would require around € 400 million/year additionally. This could save some 4.1 Mt CO_2-eq. GHG emissions/year and € 1.33 billion/year in energy costs, creating around 37 000 new jobs by 2020. With a time-span of 10 years, this would require subsidies of around € 14 billion and would permanently save in the region of 3 400 t CO_2-eq./year of emissions (Volume 3, Chapter 5).

Options to further improve energetic building renovation include increased energetic and ecological orientation of building regulations for new buildings and renovations and a shift of housing subsidies toward renovation. The quality of data on the building stock and energy consumption (particularly for commercial buildings) in Austria could be improved.

Only few studies are available on urban climate in Austria (city planning, surface coloring, greening buildings), so that an estimation of temperature reduction measures to mitigate climate-induced warming in urban areas (and heat islands) and the related energy and emission savings is not yet possible. Detailed economic studies on broad cost/benefit analyses of high-value building renovation are also lacking, as most studies focus only on individual buildings (Volume 3, Chapter 5).

S.3.9   Transformative Pathways

Without measures to curtail emissions, significant negative consequences for the biosphere and socioeconomic conditions can be expected globally. An important target value to limit

"dangerous" climate change in the sense of the United Nations Framework Convention on Climate Change (UNFCCC) is to stabilize the increase in global warming at 2 °C. In addition to vigorous mitigation, measures will also be needed to adapt to climate change that can no longer be avoided even at this low stabilization level (Volume 3, Chapter 1; Volume 3, Chapter 6).

The temperature change at the end of the 21st century and beyond depends on the amount of CO_2 emissions accumulated by then. Figure S.3.13 illustrates this correlation on the basis of several models for each of the four "representative concentration paths" (RCPs) developed for the IPCC (2013) until 2100.

Climate protection measures implemented to date have proven to be inadequate to reverse dangerous climate change. Each additional delay further decreases the chances of reaching the 2 °C stabilization target. Most measures suggested to date are "top-down" and relate to nation states. Some of these are included in international agreements. A major reason for the ineffectiveness of current climate policy is that it does not recognize what a large number of actors have a share in climate responsibility and that consequently an interactive (bottom-up and top-down), political process with feedback loops would be necessary for effective regulation. Further reasons for policy failure are the complex connections between social, economic, and environmental problems. The repeated disappointment of high expectations with regard to international climate negotiations has resulted in the disillusionment of both politicians and the public in climate politics (Volume 3, Chapter 6).

To develop viable paths to reach the 2 °C target, an understanding of the connection between environmental degradation, poverty, and social inequality is necessary. Examples of such interactions are the connections between climate change, mobility behavior, and land use change, changes in population, the health status of the population and environmental damage, technological change and global market integration, and the fact that some parts of the world are changing very rapidly, whilst others stagnate and remain in poverty (Volume 3, Chapter 6).

From a structural point of view, the climate change crisis and excessive resource use are closely related to the currently dominant economic order. From this perspective resource-intensive way of life and production modes, the fact that few govern over many, and increasing economic inequality are all both part and root causes of the climate crisis. **As currently prevailing structures and practices are responsible for the sustainability crisis, they need to be changed to overcome**

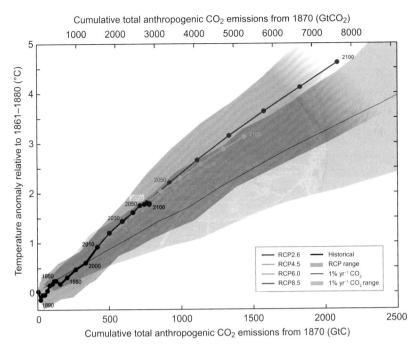

Figure S.3.13. The impact of cumulative total CO_2 emissions on temperature increases for the historic period from 1870 to 2010 and for the future using four "Representative Concentration Pathways" (RCPs). Each RCP is depicted as a coloured line, with points indicating mean decadal values. Results from empirical studies for the historical period (1860 to 2010) are indicated in black. The thin black line depicts model results with a CO_2 increase of 1% per year. The pink coloured plume illustrates the spread of the suite of ensemble models for the four RCP scenarios (see Volume 1, Chapter 1 and Volume 3, Chapter 1). These are named after their radiation forcing reached in 2100 (between 2.6 and 8.5 W/m²); Source: IPCC AR5 WG1 SPM (2013)

the crisis. Such comprehensive change processes aimed at sustainability are described as **social-ecological transformation** (Volume 3, Chapter 6). New paths and practices include transformative approaches to climate mitigation and adaptation that go beyond marginal and incremental steps. Such measures can require changes in form and structure and in so doing open up fundamentally new courses of action (Volume 3, Chapter 6).

In this sense, the earlier sectoral analysis showed that significant emission reduction potentials exist in all sectors in Austria and that measures to use these are known. However, the analysis also clearly shows that neither **currently planned, nor more sectoral, mostly technology-oriented, measures will suffice to achieve the expected Austrian contribution to the global 2°C stabilization target.** Meeting the 2°C target requires more than the implementation of incremental improvements to production technologies, greener consumer goods, and a policy that (marginally) increases efficiency in Austria. **A transformation of the interaction between economy, society, and the environment is required that is supported by behavioral changes on the part of individuals that, in turn, support such a transformation.** If the risk of unwanted, irreversible change is not to increase, the transformation needs to be

introduced and implemented rapidly (Volume 3, Chapter 6).

The aims for the development of renewable energy and energy efficiency included in the Austrian energy strategy are aligned with the EU targets for 2020, which aim for an EU-wide reduction in emissions by 20% in relation to 1990 levels. On the basis of various global climate protection scenarios, there are significant doubts as to whether the EU 2020 reduction targets are sufficient to reach the long-term goal of stabilizing temperature change at under 2°C in a cost-efficient manner (Volume 3, Chapter 1). In contrast, more stringent targets for industrialized nations in the minus 25% to minus 40% range by 2020 are being discussed, targets which are also implied by the illustrative reduction pathways in the EU "Roadmap for moving to a low-carbon economy in 2050."

When applied to Austria, EU 2020 targets are currently interpreted as a reduction commitment of roughly 3% in relation to 1990. This target for 2020 is considerably lower than the original Austrian target during the first Kyoto period for 2008–2012. As Austria is a relatively wealthy country with significant potential for renewable energy, it would be possible for Austria to at least align its climate protection targets for 2020 with the original Kyoto targets (–13% emissions in relation to 1990).

Furthermore, studies on the impacts of the economic crisis from 2008 to 2010 on the EU conclude that the crisis has made reaching the EU 2020 target of –20 % GHG emissions considerably easier than originally assumed. With additional efforts these could realistically be surpassed (Volume 3, Chapter 6).

In many policy areas, discussions on social-ecological transformation are reduced to concepts such as **"sustainable growth", "qualitative growth",** or the current variation, **"green growth".** However, these are concepts that would make production more environmentally friendly through new technologies, but leave the logic of production and consumption unchanged. In essence, "green growth" suggests a continuation of current policy measures to support economic growth, and merely enhancing these with (mostly unquantified) environmental measures. The recently published "European Report on Development" (2013) accepts green growth as a policy option, but at the same time demands a broad range of objectives and structural changes, which would allow inclusive and sustainable development at the local, national, and global levels.

Modern economies and economic research are structurally closely connected to the paradigm of unlimited economic growth, measured on the basis of Gross Domestic Product (GDP). National and international climate policy concentrates on growth-dependent policy measures. **Only a small number of studies have critically questioned the effects of stringent climate protection targets on the development paths of economies and the expected feedbacks.**

As green growth is a contentious approach, the question remains as to how climate protection and other social-ecological objectives can be achieved concurrently. For planning and political decisions and in order to steer social-ecological systems toward sustainability, **appropriate indicator systems,** which **can measure societal progress and well-being** are necessary. Several factors that contribute to quality of life, such as residential building activities, a healthy diet, health care, education, or security correlate positively with GDP, the prevalent indicator. On the other hand, factors and activities that negatively contribute to the common good, such as natural disasters, increasing environmental damage, or processes of social disintegration can also contribute to GDP growth. For this reason, a search is under way at the European and international levels for more appropriate indicators.

Taken alone, climate friendliness is a necessary, but insufficient condition for sustainable development. Achieving the 2 °C target requires there to be a simultaneous focus on climate-friendly technologies, behaviors, and institutional change. This applies particularly to energy supply and demand, industrial processes, and agriculture. These three areas are particularly important: in 2010 they caused 79 % of greenhouse gas emissions in Austria, one-third of which was caused by road traffic, 13 % by industrial processes, and 9 % by agricultural emissions. To limit the danger of irreversible damage, climate friendliness needs to be integrated into future investment, production, consumption, and political decisions as a matter of course. At the same time it is important to ensure that neither social nor economic framework conditions are overburdened. Climate friendliness needs to be integrated into the context of the significantly broader criteria of sustainability.

Although **climate friendly measures** are often connected to costs or unwanted changes, **they can cause various positive side effects,** for instance, on quality of life, health, employment, rural development and environmental protection, security of supply, and trade balance. The internalization of these positive side effects of climate protection creates the necessary room for maneuver.

There are several empirical studies that have analyzed changes in the energy system up to 2050 in Austria. They all see potential for reducing energy end use by around 50 % by 2050 (Figure S.3.14).

The energy models, with which the scenario analyses depicted in Figure S.3.14 were carried out, show that empirical studies focus mainly on changes in energy supply, **while the significant challenge of analyzing demand and energy use is for the most part not considered.** Investigating these aspects would require a significantly higher number of technical details, actors, and institutional arrangements to be considered as well as the driving forces of increasing energy demand. Nonetheless, such analyses are necessary to describe the main actors, measures, barriers, risks, and costs of transformation. As the reorganization toward climate compatibility leads not only to burdens, but can also stimulate important growth sectors, it is in both the public and economic interest to raise awareness about the new possibilities and expected redistribution processes. This is necessary to shape effective markets and, not least, to identify room for maneuver in the international negotiation of the global 2 °C target.

To assess alternative paths toward a transformation to a climate-friendly and sustainable society within the energy scenarios presented above, consideration is needed of **the effects of global and regional development dynamics,** which form the broader context for development options in Austria, and which have not been fully considered in models. In accordance with a systemic approach, a chosen framework for climate responsibility must be specified, before the possible courses of

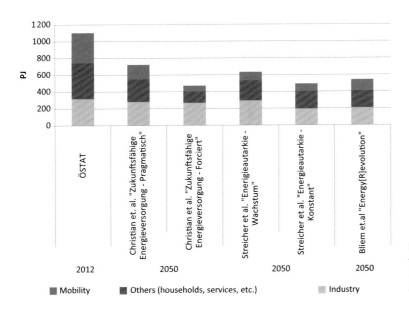

Figure S.3.14 Comparison of end energy use per sector in 2012 compared to 2050 for various scenarios; Sources: based on Statistik Austria (2011). Statistik der Zivilluftfahrt (2010)

action of individual actors are addressed, as this plays a fundamental role in defining what constitutes a room for maneuver and effects of climate protection actions.

In many respects, climate change will have larger impacts on other regions of the world, which in turn will increase migration pressures on Europe (and Austria). Although for the most part migration has taken place within world regions until now, the current flow of refugees particularly from Africa to Europe could increase in future. Changes in migration flows can result from both the impacts of extreme weather events and long-term climate variability and can be an effective means of adapting to climate change for many people.

As a small, diversified and economically open economy, Austria is open to a number of internal and external dynamics, which have been insufficiently represented in energy and emissions models to date. An example is the rapidly increasing European and global market integration and globalization that are leading to the internationalization of and an increase in complexity of process chains of processing industries and increasing geographical distances between the production and consumption of goods. Furthermore, as mentioned above, in Austria the production of imported goods causes more emissions internationally than the production of goods for export cause nationally (Volume 3, Chapter 6). Climate protection measures must consider such contexts, as scopes that are too narrow can lead to a further outsourcing of emissions and consequently fail to achieve their objective of achieving a global reduction in GHGs.

Austria has pledged to take action in the framework of EU climate policy (Volume 3, Chapter 1; Volume 3, Chapter 6). This requires a **more permanent and long-term planning of**

climate goals to be pursued, which recent Austrian climate policy has failed to achieve. Long-term, binding climate targets minimize investment risks and allow economic actors to take foresightful planning decisions for long-living infrastructure. A fundamental policy measure would be a comprehensive **evaluation of subsidies and grants for possible climate effects**, as these are important means of political management. This would apply particularly to, for example, the low petroleum tax in Austria compared with other EU countries, commuter tax breaks, residential building subsidies not connected to energy efficiency requirements, and tax breaks for air travel and company cars.

New incentive schemes that directly influence action are particularly important; they can lead to the development of new business models and slow down energy demand. Energy service companies (ESCOs) are an example of such a business model. ESCOs dispose of funds, either alone or in connection with a financial institute, from which capital funds can be taken to improve the energy efficiency of installations or buildings. Parts of the resulting savings in energy costs are then used to repay the funds used for the investment.

Attaching a cost to CO_2 can systematically steer production, consumption, and investment decisions toward climate compatibility and accelerate the decarbonization of the energy system as well as climate-compatible development (Volume 3, Chapter 6). As buyers need to pay for goods and services in proportion to their climate impact, accounting for CO_2 gives an incentive to change to alternatives with fewer climate impacts – and an incentive for producers, to reduce the carbon footprint of the goods and services they produce. This is the idea at the heart of the European Union Emissions Trading

System (EU-ETS; Volume 3, Chapter 1; Volume 3, Chapter 6). The weaknesses of the current design of the EU-ETS are the lack of adaptability of the cap and the over-allocation of certificates (leading to low prices, Volume 3, Chapter 6). As such, measures to reform the EU-ETS that would stabilize the price signal required for transformative investments would be constructive (Volume 3, Chapter 6). The options in this case are directly steering the prices or steering the quantities, as has been suggested by the EU Commission ("Market Stability Reserve"). In a number of cases the introduction of taxes on emissions has been shown to reduce emissions (Volume 3, Chapter 6).

Participatory planning processes will play an important role in the transformation toward a climate- compatible energy infrastructure. It will eventually be necessary to define new roles for individuals, networks, and communities, in order to engage on new paths toward sustainability. Communal energy networks have a long history in Austria, but are the exception rather than the rule in the current market structure. They are essential for creating new and decentralized energy technologies and the required power networks in such a way that is locally accepted.

In this context, the role of social and technological innovation will play a central role. Experimentation and learning from experience is necessary as also is the willingness to take risks and accept the fact that certain innovations will fail. This is problematic for individual companies but also for the public policy area where failure has consistently negative connotations (Volume 3, Chapter 6).

Fundamental renewal will be necessary, with respect to the goods and services that are produced in Austria, as well as to large-scale investment programs. The assessment of new technologies and social developments will need to be based on a variety of criteria (multi-criteria approach) and require integrative socio-ecologically oriented decision making, instead of short-term, narrowly defined cost-benefit calculations. Furthermore, national action should be concerted internationally, with both neighboring EU member states and the international state community and in particular in partnerships with developing countries (e.g., through cooperation in the area of technology transfer, such as the "Sustainable Energy for All" initiative).

In Austria, changes in people's belief-systems relating to sustainability can be noted and action on a local scale observed. Individual pioneers of change are already taking climate-friendly action and have developed novel business models (e.g., energy service companies in real estate, climate-friendly mobility, or local supply) and are also transforming

municipalities and regions. Climate-friendly transformation approaches can also be identified on the political level. If Austria wishes to contribute to achieving the global 2 °C target and help shape future climate-friendly development on the European and international levels, such initiatives need to be reinforced and supported by accompanying policy measures that create a reliable regulatory landscape.

Policy initiatives in climate mitigation and adaptation are necessary at all levels in Austria if the above objectives are to be achieved: at the federal and provincial levels and the level of local communities. Within the Austrian federal structure competences are split, such that only a common and mutually adjusted approach across these levels can ensure highest possible effectiveness and achievement of objectives. To effectively implement the substantial transformation that is necessary, a broad spectrum of instruments also needs to be implemented.

S.4   Figure Credits Synthesis

Figure S.1.1 IPCC, 2001: In: Climate Change 2001: The Scientific Basis. Contribution of Working Group I to the Third Assessment Report of the Intergovernmental Panel on Climate Change. Cambridge University Press, Cambridge.

Figure S.1.2 Morice, C.P., Kennedy, J.J., Rayner, N.A., Jones, P.D., 2012. Quantifying uncertainties in global and regional temperature change using an ensemble of observational estimates: The HadCRUT4 data set. J. Geophys. Res. D08101. doi:10.1029/2011JD017187

Figure S.1.3 Rogelj J, Meinshausen M, Knutti R, 2012. Global warming under old and new scenarios using IPCC climate sensitivity range estimates. Nature Clim. Change 2:248-253.

Figure S.1.4 Umweltbundesamt, 2012: Austria's National Inventory Report 2012. Submission under the United Nations Framework Convention on Climate Change and under the Kyoto Protocol. Reports, Band 0381, Wien. ISBN: 978-3-99004-184-0

Figure S.1.5 Böhm, R., Auer, I., Schöner, W., 2011. Labor über den Wolken: die Geschichte des Sonnblick-Observatoriums. Böhlau Verlag.

Figure S.1.6 Issued for the AAR14, adapted from: Kasper, A., Puxbaum, H., 1998. Seasonal variation of SO_2, HNO_3, NH3 and selected aerosol components at Sonnblick (3106 m a. s. l.). Atmospheric Environment 32, 3925–3939. doi:10.1016/S1352-2310(97)00031-9; Sanchez-Ochoa, A. and A. Kasper-Giebl, 2005: Backgroundmessungen Sonnblick. Erfassung von Gasen, Aerosol und nasser Deposition an der Hintergrundmeßstelle Sonnblick. Endbericht zum Auftrag GZ 30.955/2-VI/A/5/02 des Bundesministeriums für Bildung Wissenschaft und Kultur, Technische Universität Wien, Österreich; Effenberger, Ch., A. Kranabetter, A. Kaiser and A. Kasper-Giebl, 2008: Aerosolmessungen am Sonnblick Observatorium – Probenahme und Analyse der PM10 Fraktion. Endbericht zum Auftrag GZ 37.500/0002-VI/4/2006 des Bundesministeriums für Bildung, Wissenschaft und Kultur, Technische Universität Wien, Österreich.

Figure S.1.7 Issued for the AAR14, adapted from: Steinhilber, F., Beer, J., Fröhlich, C., 2009. Total solar irradiance during the Holocene. Geophysical Research Letters 36. doi:10.1029/2009GL040142; Vinther, B.M., Buchardt, S. L., Clausen, H.B., Dahl-Jensen, D., Johnsen, S. J., Fisher, D.A., Koerner, R.M., Raynaud, D., Lipenkov, V., Andersen, K.K., Blunier, T., Rasmussen, S. O., Steffensen, J.P., Svensson, A.M., 2009. Holocene thinning of the Greenland ice sheet. Nature 461, 385–388. doi:10.1038/nature08355; Renssen, H., Seppä, H., Heiri, O., Roche, D.M., Goosse, H., Fichefet, T., 2009. The spatial and temporal complexity of the Holocene thermal maximum. Nature Geoscience 2, 411–414. doi:10.1038/ngeo513; Hormes, A., Müller, B.U., Schlüchter, C., 2001. The Alps with little ice: evidence for eight Holocene phases of reduced glacier extent in the Central Swiss Alps. The Holocene 11, 255–265. doi:10.1191/095968301675275728; Nicolussi, K., Patzelt, G., 2001. Untersuchungen zur holozänen Gletscherentwicklung von Pasterze und Gepatschferner (Ostalpen). Zeitschrift für Gletscherkunde und Glazialgeologie 36, 1–87.; Joerin, U.E., Stocker, T.F., Schlüchter, C., 2006. Multicentury glacier fluctuations in the Swiss Alps during the Holocene. The Holocene 16, 697–704. doi:10.1191/0959683606hl964rp; Joerin, U.E., Nicolussi, K., Fischer, A., Stocker, T.F., Schlüchter, C., 2008. Holocene optimum events inferred from subglacial sediments at Tschierva Glacier, Eastern Swiss Alps. Quaternary Science Reviews 27, 337–350. doi:10.1016/j.quascirev.2007.10.016; Drescher-Schneider, R., Kellerer-Pirklbauer, A., 2008. Gletscherschwund einst und heute – Neue Ergebnisse zur holozänen Vegetations- und Gletschergeschichte der Pasterze (Hohe Tauern, Österreich). Abhandlungen der Geologischen Bundesanstalt 62, 45–51.; Nicolussi, K., 2009b. Alpine Dendrochronologie – Untersuchungen zur Kenntnis der holozänen Umwelt- und Klimaentwicklung, in: Schmidt, R., Matulla, C., Psenner, R. (Eds.), Klimawandel in Österreich: Die Letzten 20.000 Jahre – und ein Blick voraus, Alpine Space – Man & Environment. Innsbruck University Press, Innsbruck, pp. 41–54.; Nicolussi, K., 2011. Gletschergeschichte der Pasterze – Spurensuche in die nacheiszeitliche Vergangenheit., in: Lieb, G.K., Slupetzky, H. (Eds.), Die Pasterze. Der Gletscher am Großglockner. Verlag Anton Pustet, pp. 24–27.; Nicolussi, K., Schlüchter, C., 2012. The 8.2 ka event—calendar-dated glacier response in the Alps. Geology 40, 819–822. doi:10.1130/ G32406.1; Nicolussi, K., Kaufmann, M., Patzelt, G., Van der Pflicht, J., Thurner, A., 2005. Holocene tree-line variability in the Kauner Valley, Central Eastern Alps, indicated by dendrochronological analysis of living trees and subfossil logs. Vegetation History and Archaeobotany 14, 221–234. doi:10.1007/s00334-005-0013-y; Heiri, O., Lotter, A.F., Hausmann, S., Kienast, F., 2003. A chironomid-based Holocene summer air temperature reconstruction from the Swiss Alps. The Holocene 13, 477–484. doi:10.1191/0959683603hl640ft; Ilyashuk, E.A., Koinig, K.A., Heiri, O., Ilyashuk, B.P., Psenner, R., 2011. Holocene temperature variations at a high-altitude site in the Eastern Alps: a chironomid record from Schwarzsee ob Sölden, Austria. Quaternary Science Reviews 30, 176–191. doi:10.1016/j. quascirev.2010.10.008; Fohlmeister, J., Vollweiler, N., Spötl, C., Mangini, A., 2013. COMNISPA II: Update of a mid-European isotope climate record, 11 ka to present. The Holocene 23, 749–754.doi:10.1177/0959683612465446; Magny, M., 2004. Holocene climate variability as reflected by mid-European lake-level fluctuations and its probable impact on prehistoric human settlements. Quaternary International 113, 65–79. doi:10.1016/S1040-6182(03)00080-6; Magny, M., Galop, D., Bellintani, P., Desmet, M., Didier, J., Haas, J.N., Martinelli, N.,

Pedrotti, A., Scandolari, R., Stock, A., Vannière, B., 2009. Late-Holocene climatic variability south of the Alps as recorded by lake-level fluctuations at Lake Ledro, Trentino, Italy. The Holocene 19, 575–589. doi:10.1177/0959683609104032

Figure S.1.8 Böhm, R., 2012. Changes of regional climate variability in central Europe during the past 250 years. The European Physical Journal Plus 127. doi:10.1140/epjp/i2012-12054-6. Data source: Auer, I., Böhm, R., Jurkovic, A., Lipa, W., Orlik, A., Potzmann, R., Schöner, W., Ungersböck, M., Matulla, C., Briffa, K., Jones, P., Efthymiadis, D., Brunetti, M., Nanni, T., Maugeri, M., Mercalli, L., Mestre, O., Moisselin, J.-M., Begert, M., Müller-Westermeier, G., Kveton, V., Bochnicek, O., Stastny, P., Lapin, M., Szalai, S. , Szentimrey, T., Cegnar, T., Dolinar, M., Gajic-Capka, M., Zaninovic, K., Majstorovic, Z., Nieplova, E., 2007. HISTALP—historical instrumental climatological surface time series of the Greater Alpine Region. International Journal of Climatology 27, 17–46. doi:10.1002/joc.1377, as well as data from: Climatic Research Unit, University of East Anglia, http://www.cru.uea.ac.uk/

Figure S.1.9 Böhm, R., 2012. Changes of regional climate variability in central Europe during the past 250 years. The European Physical Journal Plus 127. doi:10.1140/epjp/i2012-12054-6; data source: Auer, I., Böhm, R., Jurkovic, A., Lipa, W., Orlik, A., Potzmann, R., Schöner, W., Ungersböck, M., Matulla, C., Briffa, K., Jones, P., Efthymiadis, D., Brunetti, M., Nanni, T., Maugeri, M., Mercalli, L., Mestre, O., Moisselin, J.-M., Begert, M., Müller-Westermeier, G., Kveton, V., Bochnicek, O., Stastny, P., Lapin, M., Szalai, S., Szentimrey, T., Cegnar, T., Dolinar, M., Gajic-Capka, M., Zaninovic, K., Majstorovic, Z., Nieplova, E., 2007. HISTALP – historical instrumental climatological surface time series of the Greater Alpine Region. International Journal of Climatology 27, 17–46. doi:10.1002/joc.1377

Figure S.1.10 Auer, I., Böhm, R., Jurkovic, A., Lipa, W., Orlik, A., Potzmann, R., Schöner, W., Ungersböck, M., Matulla, C., Briffa, K., Jones, P., Efthymiadis, D., Brunetti, M., Nanni, T., Maugeri, M., Mercalli, L., Mestre, O., Moisselin, J.-M., Begert, M., Müller-Westermeier, G., Kveton, V., Bochnicek, O., Stastny, P., Lapin, M., Szalai, S., Szentimrey, T., Cegnar, T., Dolinar, M., Gajic-Capka, M., Zaninovic, K., Majstorovic, Z., Nieplova, E., 2007. HISTALP—historical instrumental climatological surface time series of the Greater Alpine Region. International Journal of Climatology 27, 17–46. doi:10.1002/joc.1377; ENSEMBLES project: Funded by the European Commission's 6th Framework Programme through contract GOCE-CT-2003-505539; reclip:century: Funded by the Austrian Climate Research Program (ACRP), Klimaund Energiefonds der Bundesregierung, Project number A760437

Figure S.1.11 Auer, I., Böhm, R., Jurkovic, A., Lipa, W., Orlik, A., Potzmann, R., Schöner, W., Ungersböck, M., Matulla, C., Briffa, K., Jones, P., Efthymiadis, D., Brunetti, M., Nanni, T., Maugeri, M., Mercalli, L., Mestre, O., Moisselin, J.-M., Begert, M., Müller-Westermeier, G., Kveton, V., Bochnicek, O., Stastny, P., Lapin, M., Szalai, S., Szentimrey, T., Cegnar, T., Dolinar, M., Gajic-Capka, M., Zaninovic, K., Majstorovic, Z., Nieplova, E., 2007. HISTALP – historical instrumental climatological surface time series of the Greater Alpine Region. International Journal of Climatology 27, 17–46. doi:10.1002/joc.1377; ENSEMBLES project: Funded by the European Commission's 6th Framework Programme through contract GOCE-CT-2003-505539; reclip:century: Funded by the Austrian Climate Research Program (ACRP), Klimaund Energiefonds der Bundesregierung, Project number A760437

Figure S.1.12 Gobiet, A., Kotlarski, S., Beniston, M., Heinrich, G., Rajczak, J., Stoffel, M., n.d. 21st century climate change in the European Alps – A review. Science of The Total Environment. doi:10.1016/j.scitotenv.2013.07.050

Figure S.2.1 Issued for the AAR14.

Figure S.2.2 Coy, M.; Stötter, J., 2013: Die Herausforderungen des Globalen Wandels. In: Borsdorf, A.: Forschen im Gebirge –Investigating the mountains – Investigando las montanas. Christoph Stadel zum 75. Geburtstag. Wien: Verlag der Österreichischen Akademie der Wissenschaften

Figure S.2.3 Dokulil, M.T., 2009: Abschätzung der klimabedingten Temperaturänderungen bis zum Jahr 2050 während der Badesaison. Bericht Österreichische Bundesforste, ÖBf AG. Available under: http://www.bundesforste.at/uploads/publikationen/Klimastudie_Seen_2009_Dokulil.pdf

Figure S.2.4 Austrian Federal Ministry of Agriculture, Forestry, Environment and Water Management; Dep. IV/4 – Water balance

Figure S.2.5 IPCC, 2007: In: Climate Change 2007: Impacts, Adaptation and Vulnerability. Working Group II Contribution to the Fourth Assessment Report of the Intergovernmental Panel on Climate Change. Cambridge University Press, Cambridge.

Figure S.2.6 Issued for the AAR14, data source: Munich Re, NatCatSERVICE 2014

Figure S.3.1 IPCC, 2013: In: Climate Change 2013: The Physical Science Basis. Contribution of Working Group I to the Fifth Assessment Report of the Intergovernmental Panel on Climate Change [Stocker, T.F., D. Qin, G.-K. Plattner, M. Tignor,S. K. Allen, J. Boschung, A. Nauels, Y. Xia, V. Bex and P.M. Midgley (eds.)]. Cambridge University Press, Cambridge, United Kingdom and New York, NY, USA.; IPCC, 2000: Special Report on Emissions Scenarios [Nebojsa Nakicenovic and Rob Swart (Eds.)]. Cambridge University Press, UK.; GEA, 2012: Global Energy Assessment - Toward a Sustainable Future, Cambridge University Press, Cambridge, UK and New York, NY, USA and the International Institute for Applied Systems Analysis, Laxenburg, Austria.

Figure S.3.2 Schleicher, Stefan P.,2014. Tracing the decline of EU GHG emissions. Impacts of structural changes of the energy system and economic activity. Policy Brief. Wegener Center for Climate and Global Change, Graz. Based on data by Eurostat

Figure S.3.3 Issued for the AAR14, based on: GLP, 2005. Global Land Project. Science Plan and Implementation Strategy. IGBP Report No. 53/IHDP Report No. 19. IGBP Secretariat, Stockholm. Available under: http://www.globallandproject.org/publications/ science_plan.php; Millennium Ecosystem Assessment, 2005. Ecosystems and Human Well-being: Synthesis. Island Press, Washington, DC. Available under: http://www.unep.org/maweb/en/Synthesis. aspx; Turner, B.L., Lambin, E.F., Reenberg, A., 2007. The emergence of land change science for global environmental change and sustainability. PNAS 104, 20666–20671. doi:10.1073/pnas. 0704119104.

Figure S.3.4 Umweltbundesamt, 2012: Austria's National Inventory Report 2012. Submission under the United Nations Framework Convention on Climate Change and under the Kyoto Protocol. Reports, Band 0381, Wien. ISBN: 978-3-99004-184-0

Figure S.3.5 Issued for the AAR14 by R. Haas based on data by Energy Economics Group and Statistics Austria, 2013a. Energiebilanzen 1970-2011 [WWW Document]. URL http://www.statistik.gv.at/web_de/statistiken/energie_und_umwelt/energie/energiebilanzen/index.html (accessed 7.14.14).

Figure S.3.6 Hausberger,S., Schwingshackl, M., 2011. Update der Emissionsprognose Verkehr Österreich bis 2030 (Studie erstellt im Auftrag des Klima- und Energiefonds No. Inst-03/11/ Haus Em 09/10-679). Technische Universität, Graz.

Figure S.3.7 translated for the AAR14 adapted from ADEME, 2007; US DoT, 2010; Der Boer et al., 2011; NTM, 2012; WBCSD, 2012, In Sims R., R. Schaeffer, F. Creutzig, X. Cruz-Núñez, M. D'Agosto, D. Dimitriu, M.J. Figueroa Meza, L. Fulton, S. Kobayashi, O. Lah, A. McKinnon, P. Newman, M. Ouyang, J.J. Schauer, D. Sperling, and G. Tiwari, 2014: Transport. In: Climate Change 2014: Mitigation of Climate Change. Contribution of Working Group III to the Fifth Assessment Report of the Intergovernmental Panel on Climate Change [Edenhofer, O., R. Pichs-Madruga, Y. Sokona, E. Farahani, S. Kadner, K. Seyboth, A. Adler, I. Baum, S. Brunner, P. Eickemeier, B. Kriemann, J. Savolainen, S. Schlömer, C. von Stechow, T. Zwickel and J.C. Minx (eds.)]. Cambridge University Press, Cambridge, United Kingdom and New York, NY, USA

Figure S.3.8 UNWTO-UNEP-WMO, 2008: Climate change and tourism – Responding to global challenges. UNWTO: Madrid, Spain. Available under: http://www.unep.fr/scp/publications/details. asp?id=WEB/0142/PA

Figure S.3.9 Issued for the AAR14, data source: STATcube – Statistische Datenbank von Statistik Austria. Available under: http://sdb.statistik.at/superwebguest/autoLoad.do?db=deeehh

Figure S.3.10 Muñoz, P., Steininger, K.W., 2010. Austria's CO$_2$ responsibility and the carbon content of its international trade. Ecological Economics 69, 2003–2019. doi:10.1016/j.ecolecon.2010.05.017

Figure S.3.11 Issued for AAR14, data source: STATcube – Statistische Datenbank von Statistik Austria. Available under: http://sdb.statistik.at/superwebguest/autoLoad.do?db=deeehh

Figure S.3.12 Issued for the AAR14, data source: Umweltbundesamt, 2009: Klimaschutzbericht 2009. Reports, Band 0226, Wien. ISBN: 978-3-99004-024-9; Umweltbundesamt, 2011: Klimaschutzbericht 2011. Reports, Band 0334, Wien. ISBN: 978-3-99004-136-9; Umweltbundesamt, 2012: Klimaschutzbericht 2012. Reports, Band 0391, Wien. ISBN: 978-3-99004-194-9

Figure S.3.13 IPCC, 2013: Summary for Policymakers. In: Climate Change 2013: The Physical Science Basis. Working Group I Contribution to the Fifth Assessment Report of the Intergovernmental Panel on Climate Change [Stocker,T.F., D.Qin, G.-K. Plattner, M.Tignor, S. K.Allen, J.Boschung, A.Nauels, Y.Xia, V.Bex and P.M. Midgley (eds.)]. Cambridge University Press, Cambridge, UK and New York, USA.

Figure S.3.14 Issued for the AAR14, data source: Statistik Austria, 2011: Statistik der Zivilluftfahrt 2010. Wien. ISBN 978-3-902791-15-3. Available under: http://www.statistik.at/web_de/dynamic/services/publikationen/14/publdetail?id=14&listid=14&detail=489; Bliem, M., B. Friedl, T. Balabanov and I. Zielinska, 2011: Energie [R]evolution 2050. Der Weg zu einer sauberen Energie-Zukunft in Österreich. Endbericht. Institut für Höhere Studien (IHS), Wien; Christian et al., 2011: Zukunfsfähige Energieversorgung für Österreich (ZEFÖ). Vienna, Umweltmanagement Austria, Institut für industrielle Ökologie und Forum Wissenschaft & Umwelt im Rahmen des Programmes „Energie der Zukunft" des BMVIT. Streicher, W., H. Schnitzer, M. Titz, F. Tatzber, R. Heimrath, I. Wetz, S. Hausberger, R. Haas, G. Kalt, A. Damm, K. Steininger and S. Oblasser, 2011: Energieautarkie für Österreich 2050. funded by the Austrian climate and energy fund (kli:en). Universität Innsbruck – Institut für Konstruktion und Materialwissenschaften, Arbeitsbereich Energieeffizientes Bauen, Innsbruck

Appendix
Underlying Documents

Appendix: Underlying Documents

Citation of the Summary for Policymakers (SPM)

APCC (2014): Summary for Policymakers (SPM), revised edition. In: Austrian Assessment Report Climate Change 2014 (AAR14), Austrian Panel on Climate Change (APCC), Austrian Academy of Sciences Press, Vienna, Austria.

Citation of the Synthesis

Kromp-Kolb, H., N. Nakicenovic, R. Seidl, K. Steininger, B. Ahrens, I. Auer, A. Baumgarten, B. Bednar-Friedl, J. Eitzinger, U. Foelsche, H. Formayer, C.Geitner, T. Glade, A. Gobiet, G. Grabherr, R. Haas, H. Haberl, L. Haimberger, R. Hitzenberger, M. König, A. Köppl, M. Lexer, W. Loibl, R. Molitor, H.Moshammer, H-P. Nachtnebel, F. Prettenthaler, W.Rabitsch, K. Radunsky, L. Schneider, H. Schnitzer, W.Schöner, N. Schulz, P. Seibert, S. Stagl, R. Steiger, H.Stötter, W. Streicher, W. Winiwarter (2014): Synthesis. In: Austrian Assessment Report Climate Change 2014 (AAR14), Austrian Panel on Climate Change (APCC), Austrian Academy of Sciences Press, Vienna, Austria.

Documents that the SPM and Synthesis build upon.

This SPM and Synthesis build upon the following detailed publications, as published in
APCC (2014): Österreichischer Sachstandsbericht Klimawandel 2014 (AAR14). Austrian Panel on Climate Change (APCC), Verlag der Österreichischen Akademie der Wissenschaften, Wien, Österreich, 1096 pages. ISBN 978-3-7001-7699-2

Volume 1: Klimawandel in Österreich: Einflussfaktoren und Ausprägungen

Haimberger, L., P. Seibert, R. Hitzenberger, A. Steiner und P. Weihs (2014): Das globale Klimasystem und Ursachen des Klimawandels. In: Österreichischer Sachstandsbericht Klimawandel 2014 (AAR14).

Austrian Panel on Climate Change (APCC), Verlag der Österreichischen Akademie der Wissenschaften, Wien, Österreich, p. 137–172.

Winiwarter, W., R. Hitzenberger, B. Amon, H. Bauer†, R. Jandl, A. Kasper-Giebl, G. Mauschitz, W. Spangl, A. Zechmeister und S. Zechmeister-Boltenstern, (2014): Emissionen und Konzentrationen von strahlungswirksamen atmosphärischen Spurenstoffen. In: Österreichischer Sachstandsbericht Klimawandel 2014 (AAR14). Austrian Panel on Climate Change (APCC), Verlag der Österreichischen Akademie der Wissenschaften, Wien, Österreich, p. 173–226.

Auer, I., U. Foelsche, R. Böhm†, B. Chimani, L. Haimberger, H. Kerschner, K.A. Koinig, K. Nicolussi und C. Spötl, 2014: Vergangene Klimaänderung in Österreich. In: Österreichischer Sachstandsbericht Klimawandel 2014 (AAR14). Austrian Panel on Climate Change (APCC), Verlag der Österreichischen Akademie der Wissenschaften, Wien, Österreich, p. 227–300.

Ahrens, B., H. Formayer, A. Gobiet, G. Heinrich, M. Hofstätter, C. Matulla, A.F. Prein und H. Truhetz, 2014: Zukünftige Klimaentwicklung. In: Österreichischer Sachstandsbericht Klimawandel 2014 (AAR14). Austrian Panel on Climate Change (APCC), Verlag der Österreichischen Akademie der Wissenschaften, Wien, Österreich, p. 301–346.

Schöner, W., A. Gobiet, H. Kromp-Kolb, R. Böhm†, M. Hofstätter und M. Zuvela-Aloise, 2014: Zusammenschau, Schlussfolgerungen und Perspektiven. In: Österreichischer Sachstandsbericht Klimawandel 2014 (AAR14). Austrian Panel on Climate Change (APCC), Verlag der Österreichischen Akademie der Wissenschaften, Wien, Österreich, p. 347–380.

Volume 2: Klimawandel in Österreich: Auswirkungen auf Umwelt und Gesellschaft

Stötter, J., H. Formayer, F. Prettenthaler, M. Coy, M. Monreal und U. Tappeiner, 2014: Zur Kopplung zwischen Treiber- und Reaktionssystemen sowie zur Bewertung von Folgen des Klimawandels. In: Österreichischer Sachstandsbericht Klimawandel

2014 (AAR14). Austrian Panel on Climate Change (APCC), Verlag der Österreichischen Akademie der Wissenschaften, Wien, Österreich, p. 383–410.

Nachtnebel, H.P., M. Dokulil, M. Kuhn, W. Loiskandl, R. Sailer, W. Schöner 2014: Der Einfluss des Klimawandels auf die Hydrosphäre. In: Österreichischer Sachstandsbericht Klimawandel 2014 (AAR14). Austrian Panel on Climate Change (APCC), Verlag der Österreichischen Akademie der Wissenschaften, Wien, Österreich, p. 411–466.

Lexer, M.J., W. Rabitsch, G. Grabherr, M. Dokulil, S. Dullinger, J. Eitzinger, M. Englisch, F. Essl, G. Gollmann, M. Gottfried, W. Graf, G. Hoch, R. Jandl, A. Kahrer, M. Kainz, T. Kirisits, S. Netherer, H. Pauli, E. Rott, C. Schleper, A. Schmidt- Kloiber, S. Schmutz, A. Schopf, R. Seidl, W. Vogl, H. Winkler, H. Zechmeister, 2014: Der Einfluss des Klimawandels auf die Biosphäre und Ökosystemleistungen. In: Österreichischer Sachstandsbericht Klimawandel 2014 (AAR14). Austrian Panel on Climate Change (APCC), Verlag der Österreichischen Akademie der Wissenschaften, Wien, Österreich, p. 467–556.

Glade, T., R. Bell, P. Dobesberger, C. Embleton-Hamann, R. Fromm, S. Fuchs, K. Hagen, J. Hübl, G. Lieb, J.C. Otto, F. Perzl, R. Peticzka, C. Prager, C. Samimi, O. Sass, W. Schöner, D. Schröter, L. Schrott, C. Zangerl und A. Zeidler, 2014: Der Einfluss des Klimawandels auf die Reliefsphäre. In: Österreichischer Sachstandsbericht Klimawandel 2014 (AAR14). Austrian Panel on Climate Change (APCC), Verlag der Österreichischen Akademie der Wissenschaften, Wien, Österreich, p. 557–600.

Baumgarten, A., C. Geitner, H.P. Haslmayr und S. Zechmeister-Boltenstern, 2014: Der Einfluss des Klimawandels auf die Pedosphäre. In: Österreichischer Sachstandsbericht Klimawandel 2014 (AAR14). Austrian Panel on Climate Change (APCC), Verlag der Österreichischen Akademie der Wissenschaften, Wien, Österreich, p. 601–640.

König, M., W. Loibl, R. Steiger, H. Aspöck, B. Bednar-Friedl, K.M. Brunner, W. Haas, K.M. Höferl, M. Huttenlau, J. Walochnik und U. Weisz, 2014. Der Einfluss des Klimawandels auf die Antroposphäre. In: Österreichischer Sachstandsbericht Klimawandel 2014 (AAR14). Austrian Panel on Climate Change (APCC), Verlag der Österreichischen Akademie der Wissenschaften, Wien, Österreich, p. 641–704.

Volume 3: Klimawandel in Österreich: Vermeidung und Anpassung

Bednar-Friedl, B., K. Radunsky, M. Balas, M. Baumann, B. Buchner, V. Gaube, W. Haas, S. Kienberger, M. König, A. Köppl, L. Kranzl, J. Matzenberger, R. Mechler, N. Nakicenovic, I. Omann, A. Prutsch, A. Scharl, K. Steininger, R. Steurer und A. Türk, 2014: Emissionsminderung und Anpassung an den Klimawandel. In: Österreichischer Sachstandsbericht Klimawandel 2014 (AAR14). Austrian Panel on Climate Change (APCC), Verlag der Österreichischen Akademie der Wissenschaften, Wien, Österreich, p. 707–770.

Eitzinger, J., H. Haberl, B. Amon, B. Blamauer, F. Essl, V. Gaube, H. Habersack, R. Jandl, A. Klik, M. Lexer, W. Rauch, U. Tappeiner und S. Zechmeister-Boltenstern, 2014: Land- und Forstwirtschaft, Wasser, Ökosysteme und Biodiversität. In: Österreichischer Sachstandsbericht Klimawandel 2014 (AAR14). Austrian Panel on Climate Change (APCC), Verlag der Österreichischen Akademie der Wissenschaften, Wien, Österreich, p. 771–856.

Haas, R., R. Molitor, A. Ajanovic, T. Brezina, M. Hartner, P. Hirschler, G. Kalt, C. Kettner, L. Kranzl, N. Kreuzinger, T. Macoun, M. Paula, G. Resch, K. Steininger, A. Türk und S. Zech, 2014: Energie und Verkehr. In: Österreichischer Sachstandsbericht Klimawandel 2014 (AAR14). Austrian Panel on Climate Change (APCC), Verlag der Österreichischen Akademie der Wissenschaften, Wien, Österreich, p. 857–932.

Moshammer, H., F. Prettenthaler, A. Damm, H.P. Hutter, A. Jiricka, J. Köberl, C. Neger, U. Pröbstl-Haider, M. Radlherr, K. Renoldner, R. Steiger, P. Wallner und C. Winkler, 2014: Gesundheit und Tourismus. In: Österreichischer Sachstandsbericht Klimawandel 2014 (AAR14). Austrian Panel on Climate Change (APCC), Verlag der Österreichischen Akademie der Wissenschaften, Wien, Österreich, p. 933–978.

Schnitzer, H., W. Streicher und K.W. Steininger, 2014: Produktion und Gebäude. In: Österreichischer Sachstandsbericht Klimawandel 2014 (AAR14). Austrian Panel on Climate Change (APCC), Verlag der Österreichischen Akademie der Wissenschaften, Wien, Österreich, p. 979–1024.

S. Stagl, Schulz, N., K. Kratena, R. Mechler, E. Pirgmaier, K. Radunsky, A. Rezai und A. Köppl, 2014: Transformationspfade. In: Österreichischer Sachstandsbericht Klimawandel 2014 (AAR14). Austrian Panel on Climate Change (APCC), Verlag der Österreichischen Akademie der Wissenschaften, Wien, Österreich, p. 1025–1076.